Water Resources: Pollution and Management

Water Resources: Pollution and Management

Edited by **Herbert Lotus**

New York

Published by Callisto Reference,
106 Park Avenue, Suite 200,
New York, NY 10016, USA
www.callistoreference.com

Water Resources: Pollution and Management
Edited by Herbert Lotus

© 2016 Callisto Reference

International Standard Book Number: 978-1-63239-761-4 (Hardback)

Printed in the United States of America.

Contents

Preface   VII

Chapter 1   **Domestic Water Supply Dynamics using Stable Isotopes δ^{18}O, δD, and d-Excess**   1
Deborah Leslie, Kathleen Welch, William Berry Lyons

Chapter 2   **Stable Isotopes Studies in the Urucu Oil Province, Amazon Region, Brazil**   17
Eliene Lopes de Souza, Paulo Galvão, Roseli de Almeida, Cleane Pinheiro,
Marcus Baessa, Marcio Cabral

Chapter 3   **Employing Water Demand Management Option for the Improvement of
Water Supply and Sanitation in Nigeria**   29
Emma E. Ezenwaji, Bede M. Eduputa, Joseph E. Ogbuozobe

Chapter 4   **Pollution of Wells' Water with Some Elements, Fe, Mn, Zn, Cu, Co, Pb, Cd, and
Nickel in Al-Jadriah District, Baghdad Government**   41
Hussein Mahmood Shukri Hussein

Chapter 5   **Development of a Strategic Planning Model for a Municipal Water Supply
Scheme using System Dynamics**   46
Adclere E. Adeniran, Olufemi A. Bamiro

Chapter 6   **Water Resources Conflict Management of Nyabarongo River and Kagera River
Watershed in Africa**   58
Telesphore Habiyakare, Nianqing Zhou

Chapter 7   **Water Pollution and Environmental Governance of the Tai and Chao Lake Basins
in China in an International Perspective**   66
Lei Qiu, Meine Pieter Van Dijk, Huimin Wang

Chapter 8   **Turkey Creek — A Case Study of Ecohydrology and Integrated Watershed
Management in the Low-Gradient Atlantic Coastal Plain, USA**   79
Devendra Amatya, Timothy Callahan, William Hansen, Carl Trettin,
Artur Radecki-Pawlik, Patrick Meire

Chapter 9   **Assessment of the Effectiveness of Watershed Management Intervention in
Chena Woreda, Kaffa Zone, Southwestern Ethiopia**   102
Yericho Berhanu Meshesha, Belay Simane Birhanu

Chapter 10   **The Modern Problems of Sustainable use and Management of Irrigated Lands
on the Example of the Bukhara Region (Uzbekistan)**   115
R. Kulmatov, A. Rasulov, D. Kulmatova, B. Rozilhodjaev, M. Groll

Chapter 11   **Fate of Nutrients, Trace Metals, Bacteria, and Pesticides in Nursery Recycled Water**   131
Yun-Ya Yang, Gurpal S. Toor

Chapter 12   **Environmental Anthropological Study of Watershed Management-Water Quality Conservation of Forest as a Catchment Area in the Southern Part of Australia**   **138**
Akira Hiratsuka, Yugo Tomonaga, Yoshiro Yasuda

Chapter 13   **Feasibility Analysis of MERIS as a Tool for Monitoring Lake Guiers (Senegal) Water Quality**   **150**
Seybatou Diop, Souléye Wade, Moshood N. Tijani

Chapter 14   **The Challenges of Water Pollution, Threat to Public Health, Flaws of Water Laws and Policies in Pakistan**   **170**
Azra Jabeen, Xisheng Huang, Muhammad Aamir

Permissions

List of Contributors

Preface

The world is advancing at a fast pace like never before. Therefore, the need is to keep up with the latest developments. This book was an idea that came to fruition when the specialists in the area realized the need to coordinate together and document essential themes in the subject. That's when I was requested to be the editor. Editing this book has been an honour as it brings together diverse authors researching on different streams of the field. The book collates essential materials contributed by veterans in the area which can be utilized by students and researchers alike.

This book provides comprehensive insights into the types of water resources and their uses. Water resources include all the sources of water present on earth. If we compare the proportion of fresh water with salty water, it appears to be negligible. Although freshwater is a renewable resource, the concentration of groundwater is depleting at much faster rate. Water pollution results in contamination of water resources. Pollutants may come from point or non-point sources which includes sewage treatment plants, factories, storm water, etc. Management of water pollution includes sewage treatment, industrial wastewater management, control of urban runoff, erosion, sediment control, etc. The topics included in this extensive book deal with the core aspects related to the pollution and management of water resources. It aims to shed light on some of the unexplored areas of this field. This text also includes contributions of experts from across the globe. Coherent flow of topics, reader-friendly language and extensive use of examples make this book an invaluable source of knowledge.

Each chapter is a sole-standing publication that reflects each author's interpretation. Thus, the book displays a multi-facetted picture of our current understanding of applications and diverse aspects of the field. I would like to thank the contributors of this book and my family for their endless support.

Editor

Domestic Water Supply Dynamics Using Stable Isotopes δ^{18}O, δD, and d-Excess

Deborah Leslie, Kathleen Welch, William Berry Lyons

School of Earth Sciences, The Ohio State University, Columbus, USA
Email: leslie.deborah@gmail.com, welch.189@osu.edu, lyons.142@osu.edu

Abstract

Surface water is the greatest contributor to many water supplies in urbanized areas. Understanding local water sources and seasonality is important in evaluating water resource management, which is essential to ensure the sustainability of water supplies to provide potable water. Here we describe the municipal water cycle of Columbus, Ohio, USA, using δ^{18}O, δD, and d-excess, and follow water from precipitation through surface reservoirs to a residential tap between May 2010 and November 2011. We show that trends in water isotopic composition of Ohio precipitation have a seasonal character with more negative values during the winter months and more positive values during the summer months. The year of 2011 was the wettest year on record in Central Ohio, with many months having high d-excess values (>+15‰), suggestive of increased moisture recycling, and possibly moisture introduced from more local sources. Tap waters experienced little lag time in the managed system, having a residence time of ~2 months in the reservoirs. Tap waters and reservoir waters preserved the isotopic signal of the precipitation, but the reservoir morphology also influenced the water residence time, and hence, the isotopic relationship to the precipitation. The reservoirs supplied by the Scioto River function like a river system with a fast throughput of water. The other reservoirs display more constant solute concentrations, longer flow-through times, and more lacustrine qualities. This work provides a basic understanding of a regional water supply system in Central Ohio and helps characterize the water flow in the system. These data will provide useful baseline information for the future as urban populations grow and the climate and hydrologic cycle changes.

Keywords

Ohio Precipitation, Municipal Water Supply, Moisture Recycling, Reservoir Residence Time

1. Introduction

Anticipation of the future relationship between water resources and climatological conditions and population growth is essential for developing strategies to ensure the long-term sustainability of water supplies [1]. Surface reservoirs provide approximately 63% of all public water supplies in the USA [2]. The stability of water resources in regards to monitoring and planning represents a major challenge for water managers. The timing of precipitation and evaporative water loss from reservoirs can significantly affect water storage, proving difficulty in developing reservoir-planning models [3] [4]. In addition, changes in precipitation amounts, recharge rates, and land use can affect the residence time of water through watersheds altering reservoir storage capacity. In Ohio, the total water withdrawals in 2000 were 42.0 Mm^3/d, with 39.0 Mm^3/d from surface water and only 3.0 Mm^3/d from groundwater [2]. The primary uses of this water are domestic supply, irrigation for agriculture, livestock usage, and aquaculture. For Ohio in 2000, the public water supply usage, water withdrawn by public and private water suppliers, was 5.6 Mm^3/d [2].

The public water supply of Columbus, Ohio, utilizes surface water from Griggs and O'Shaughnessy Reservoirs on the Scioto River, Hoover Reservoir on Big Walnut Creek, supplemental water from Alum Creek Reservoir on a tributary of Big Walnut Creek, and groundwater. This mix of water is distributed to residents via three water plants, Dublin Road Water Plant (DRWP), Hap Cremean Water Plant (HCWP), and Parsons Avenue Water Plant (PAWP) (**Figure 1**). With increasing suburbanization in nearby Franklin and Delaware counties and population growth in the city of Columbus, the need to understand current water resource dynamics is extremely important. Our goal in this paper is to provide data that documents the current status and stability of regional water supply so this information can serve as a baseline in the future.

Stable isotope analyses can provide an important method in determining the sources and residence times of water, monitoring the rates of water loss, and regional water resource sensitivity to evaporation. Although, the

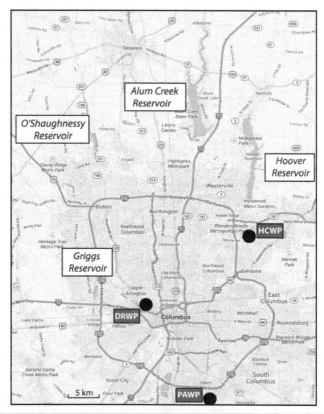

Figure 1. Reservoirs that contribute to the Columbus, Ohio water supply are Griggs, O'Shaughnessy, Alum Creek, and Hoover. The Dublin Road Water Plant (DRWP) extracts water from Griggs and O'Shaughnessy for water supply usage, while the Hap Cremean Water Plant (HCWP) utilizes water from Hoover Reservoir. Alum Creek Reservoir provides supplemental water to Hap Cremean Water Plant (adapted from Google maps). Parsons Avenue Water Plant (PAWP) distributes groundwater from wells.

isotopic evidence for evaporation does not in itself provide a warning signal of water resource sensitivity, data collected over time and analyzed in combination with information on regional climate and hydrology could be used to characterize and monitor surface water resource susceptibility to climate change [1]. The comparison of the isotopic signature of precipitation to potable water can also provide important information regarding the residence time of water in human-dominated hydrologic systems. At a regional level, the difference in isotopic ratios between precipitation and tap water can be attributed to hydrological factors such as water transit times from the source to the consumer. Management of water resources requires that the human interactions, known perturbations, and natural processes in the hydrologic cycle be fully understood. This information will enhance knowledge about site-specific hydrology, water management, water supply infrastructures, and regional hydroclimatological impacts [1]. This work utilizes stable isotopes $\delta^{18}O$ and δD to delineate the residence time of water in a human-controlled watershed-reservoir system, based on the lag between precipitation and residential tap water over time. Our observations will contribute an important dataset for interpreting regional water sources and supplies in order to evaluate water management. A major justification for this study is to provide the initial database for future comparisons of the water distribution system in Columbus, Ohio.

2. Methods

2.1. History of Columbus, Ohio, Water Supply

In 2010, Columbus, Ohio, was the 15th largest city in the United States with a population of 787,033 [5]. Water availability played a crucial role in population growth of Columbus over the past 100 years. In 1904, the city constructed Griggs Dam on the Scioto River to provide an adequate water supply. For 20 years, Griggs Reservoir served as the only reservoir providing water to Columbus. With the completion of the O'Shaughnessy Dam, an additional storage dam further upstream on the Scioto River from Griggs Reservoir, in 1925, Columbus had a water supply to serve a population of 0.5 million—twice the city's size. By 1945, population growth increased the demand for water. In 1955, Hoover Dam was completed on Big Walnut Creek. By the late 1960s, additional water supply was needed, and Alum Creek Reservoir was built in 1978, as a supplement to Hoover Reservoir. In 1983, wells in Southern Franklin County Ohio were constructed in an aquifer between the Scioto River and Big Walnut Creek. Currently three reservoirs, Griggs, O'Shaughnessy, and Hoover, provide 85% of the more than 4.9×10^5 m^3 of daily water supplied to the metropolitan area, while the remaining 15% is drawn from wells in southern Franklin County, Ohio (**Table 1**) [6].

2.2. Site Description-Reservoirs and Water Plants

Griggs Reservoir (**Figure 1**) is associated with the Scioto River (which remains the main water source for Columbus) (**Table 1**). It is ~9.7 km long, and located on the NW side of Columbus, Ohio. Approximately 16 km upstream of Griggs Dam on the Scioto River is the O'Shaughnessy Dam separating the slightly larger O'Shaughnessy Reservoir (**Figure 1**) from Griggs Reservoir near Dublin, Ohio. Together, these two reservoirs have a capacity of 2.3×10^7 m^3 [6]. Approximately 30% of drinking water for the city of Columbus comes from the Griggs and O'Shaughnessy Reservoirs. The Griggs and O'Shaughnessy Reservoirs retain only a small percentage of the Scioto River water that flows through Columbus. Griggs and O'Shaughnessy reservoirs are more of a riverine system connected to the Scioto River, and this creates a very weather- and flow-dependent system [8].

Table 1. Columbus, Ohio, water supply information.

Reservoirs/Wells	Water Source	Surface Area (km^2)	Capacity (m^3)	Average Raw Water Demand*
Griggs	Scioto River	1.5	5.3×10^6	6.4×10^4
O'Shaughnessy	Scioto River	3.4	1.8×10^7	1.1×10^5
Hoover	Big Walnut Creek	13.2	7.9×10^8	3.8×10^5
Alum Creek Lake	Alum/Big Walnut Creek	13.7	8.9×10^8	5.0×10^5
South Well-Field	Groundwater	na	na	5.7×10^5

*Average raw water demands assume 6% loss through treatment plant [7].

Hoover Dam (**Figure 1**) forms the Hoover Memorial Reservoir, which supplies water for the entire northeast portion of Franklin County (**Table 1**). Alum Creek Dam is located on Alum Creek, a tributary of Big Walnut Creek that drains into the Scioto River south of Columbus (**Table 1**). Hoover Reservoir can be supplemented with Alum Creek Reservoir water. The maximum amount allowed to be transferred from the Alum Creek Reservoir to Hoover is an average of 1.2×10^6 m^3/d per calendar year [7]. Hoover and Alum Creek reservoirs are more characteristic of a lacustrine environment, resulting in a more classic reservoir system [8].

Columbus Division of Water manages water delivered to residents of the Greater Columbus Area. Three water plants, Dublin Road Water Plant, Hap Cremean Water Plant, and Parsons Avenue Water Plant (**Figure 1**), distribute water. The Dublin Road Water Plant utilizes surface water from the Griggs and O'Shaughnessy Reservoirs on the Scioto River, and provides water to downtown Columbus, western, and southwestern Franklin County. The Dublin Road Water Plant is designed to treat up to 2.5×10^5 m^3/d. The Hap Cremean Water Plantutilizes surface water from the Hoover Reservoir on Big Walnut Creek and serves water to The Ohio State University and the northern half of Franklin County. The Hap Cremean Water Plant is designed to treat up to 4.7×10^5 m^3/d of surface water. Alum Creek Reservoir also provides supplemental water to the Hap Cremean Water Plant. The Parsons Avenue Water Plantutilizes groundwater from wells and provides water to southeastern Franklin County, Ohio. The PAWP is designed to treat up to 1.9×10^5 m^3/d of groundwater.

2.3. Sample Collection and Storage

Precipitation was collected from rain events using a precipitation collector located on the Ohio State University campus (latitude: 40.0025; longitude: -83.0390). This collector consisted of a plastic funnel (top diameter = 248 mm; height = 319 mm), silicone tubing, and a 1000 mL Nalgene® high-density polyethylene (HDPE) bottle with two fittings in the cap to allow the displaced air out as the bottle is filled with precipitation. The sample collection tubing line was contorted with a loop near the funnel to help minimize evaporation between the funnel and sample bottle. Each precipitation sample was collected once daily during events and as soon as possible after each event. Precipitation aliquots were collected for stable isotope and anion analysis, and samples were not filtered. Samples for isotopic analysis were collected in 20 mL glass scintillation vials with polyethylene cone shaped liners in the caps to help eliminate any headspace. Stable isotopic water samples were filled to the top to prevent evaporation, and lids were wrapped in parafilm to prevent any leakage or evaporative loss. All samples were then stored dark and chilled until time for analysis, usually within 2 - 3 months of collection.

Reservoir water collection was done weekly at Griggs Reservoir, and all other reservoirs were sampled on a monthly basis (May 2010-November 2011) by an individual wearing vinyl gloves. Water was collected from the reservoir pier. Reservoir samples were syringe filtered in the field using 0.4 μM Whatman™ Nucleopore syringe filters into a 60 mL pre-cleaned Nalgene® low density polyethylene (LDPE) bottle into glass 20 mL scintillation vials for stable isotopic analysis. For the isotopic samples, scintillation vial caps were wrapped in parafilm. Samples were refrigerated and stored dark until analysis.

Tap waters were collected from a residential tap in downtown Columbus supplied by the Dublin Road Water Plant weekly over the study from May 2010 to November 2011. Hence, the water collected originated from the Griggs/O'Shaughnessy reservoir system. Water from the tap was allowed to flow for ~5 seconds, and then aliquots for stable isotopes and anion analyses were collected and stored in the same way as precipitation samples.

2.4. δ^{18}O and δD Analysis

Stable isotopic samples were analyzed for δ^{18}O and δD using a Picarro Wavelength Scanned-Cavity Ring Down Spectroscopy Analyzer for Isotopic Water-Model L1102-i within 2 - 3 months of collection [9]. The stable isotopic analysis includes a calibration using internal laboratory standards of waters spanning the isotopic range of the samples. Our internal laboratory standards were deionized water aliquots from Florida (δ^{18}O = -2.3‰; δD = -12.3‰), Ohio (δ^{18}O = -9.3‰; δD = -64.6‰), Nevada (δ^{18}O = -14.2‰; δD = -106.0‰) and Colorado (δ^{18}O = -17.0‰; δD = -129.0‰). Our internal laboratory standards were standardized relative to Standard Light Antarctic Precipitation (SLAP) and Vienna Standard Mean Ocean Water (VSMOW), at the Ice Core Paleoclimatology Laboratory at The Ohio State University using standard gas-source mass spectrometry techniques [10]. Isotopic values are presented as per mil (‰) values. The accuracy was ≤4.2% for δD and ≤3.4% for δ^{18}O, calculated by relative standard error comparing the Picarro isotopic measurement for an internal laboratory standard and the Ice Core Paleoclimatology Laboratory isotopic measurement for an internal laboratory standard. Precision was calculated as ≤1.0% for δD and ≤0.5% for δ^{18}O, for 100 measurements using relative standard deviation.

Deuterium excess (d-excess) values were calculated as d-excess = δD − 8*δ^{18}O, using the Global Meteoric Water Line [11]. Deuterium excess reflects non-equilibrium fractionation characterized by temperature and relative humidity in the precipitation source region [12] [13].

2.5. Weather Station Data Collection

Temperature and precipitation amount data were gathered at the Waterman Farm of Ohio State campus weather station within the network of Ohio Agricultural Research and Development Center (OARDC) of the Ohio State University [14]. This station is ~1.4 km from the precipitation collector, and it has been collecting data since January 1986. This long-term record was used to calculate the average 25-yr monthly temperatures and precipitation amounts. Data from this study were also compared to the long-term climate records maintained by the National Weather Service.

3. Results

3.1. Meteorology

The precipitation totals for this study were above the 25-yr average monthly precipitation totals for Central Ohio for more than half of the months in this study (May 2010 to November 2011, **Figure 2**). Monthly mean temperatures were slightly above normal compared to the 25-yr monthly average temperature (**Figure 2**). Spring 2011 (February, March, April, and May) had precipitation totals much greater (as much as 2× the 25-yr average). Fall 2011 (September, October, and November) also had greater amounts of precipitation than the 25-yr monthly average. April 2011 was the wettest April on record, with total precipitation of 18.1 cm, breaking the record of 18.0 cm set in 1893. The wettest year on record in Columbus was 2011 (139.6 cm). The previous record was 135.0 cm in 1990. Columbus also experienced the third wettest spring (44.8 cm, with previous record of 48.8 cm in 1882), and the second wettest fall on record (38.1 cm, with the previous record of 39.4 cm in 1881) [15].

3.2. Isotopic Composition of Precipitation

Values of δ^{18}O and δD for precipitation ranged from −19.5‰ to −0.23‰ and from −148‰ to +4.5‰, respectively (n = 119; [16]). These ranges are typical of mid-continental stations globally [17]. The δ^{18}O volume-weighted precipitation values followed a similar trend to the precipitation δ^{18}O monthly average, with a more

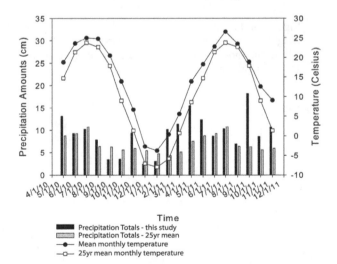

Figure 2. Monthly precipitation totals and average monthly air temperatures across the study during May 2010 to November 2011 in comparison to the 25-yr monthly precipitation totals and mean monthly air temperatures. Data gathered from a weather station within the network of OARDC.

positive signature in the spring and summer seasons and more negative signature in the fall and winter seasons (**Figure 3**). The δ^{18}O of precipitation displayed a significant relationship with the δ^{18}O monthly volume-weighted and 25-yr monthly mean temperature (r^2 = 0.66; p < 0.05). However, no statistically significant relationship existed between the stable isotopic value and the amount of precipitation per event throughout the entire data series (**Figure 3**). Although in the fall of 2011 (e.g. the 2nd wettest fall on record), precipitation values display an "amount effect" with more negative values with larger events (**Figure 3**). Individual precipitation samples span ~20‰ range for δ^{18}O and ~150‰ range for δD, with as expected more negative values from winter precipitation and more positive values from summer precipitation. The air temperature and δ^{18}O relationship form a linear regression of δ^{18}O = 0.27*T − 11.3 with r^2 = 0.44. Precipitation d-excess values spanned +0.1‰ to +29‰, and had a positive significant relationship with δD (r^2 = 0.23). The majority of data fall along or above the Global Meteoric Water Line (GMWL) (δD = 8.0*δ^{18}O + 10; [11]) (**Figure 4**). These data have a linear regression of δD = 8.04*δ^{18}O + 15.64 (r^2 = 0.96; p < 0.001) (**Figure 4**), suggesting minimal evaporation has occurred.

3.3. Reservoir Waters

The reservoir data also display a seasonal pattern, with an isotopic transition from more positive waters during

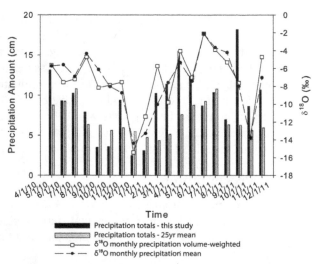

Figure 3. Precipitation δ^{18}O monthly mean and volume-weighted value from May 2010 to November 2011, in relation to the monthly precipitation amount totals compared to 25-yr monthly precipitation amount mean.

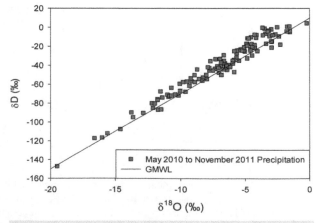

Figure 4. Precipitation isotopic values of May 2010 to November 2011 plotted in relation to the GMWL.

the warmer periods to more negative values during cooler periods (**Figure 5**). Reservoir isotopic values ranged from $\delta^{18}O$ of −8.9‰ to −4.7‰ and δD of −62‰ to −18‰, with d-excess values from +1.4‰ to +23‰ [16]. The most isotopically positive reservoir waters in 2010 occurred ~September 2010, while the 2011 seasonal enrichment occurred earlier, during August. The most negative signatures in the reservoirs were a $\delta^{18}O$ of −8.9‰ in March 2011, and in May 2010 in Hoover Reservoir. Reservoir waters follow along the GMWL, within the same range of the precipitation samples. The majority of these data fall above the line but some fall below ($\delta D = 6.0*\delta^{18}O − 1.9$; $r^2 = 0.64$; **Figure 6**). In all reservoirs, the $\delta^{18}O$-δD values displayed a positive significant correlation.

3.4. Residential Tap Waters

Residential tap water distributed via the Dublin Road water plant ranged from $\delta^{18}O$ of −9.0‰ to −4.3‰ and δD

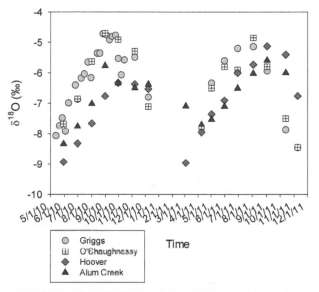

Figure 5. Reservoir $\delta^{18}O$ values from Griggs, O'Shaughnessy, Hoover, and Alum Creek from May 2010 to November 2011.

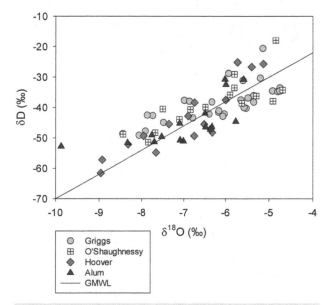

Figure 6. Reservoirs isotopic values over May 2010 to November 2011 plotted in relation to the GMWL.

of −59‰ to −19‰, respectively (n = 53; **Figure 7**; [16]). D-excess values range from +1.2‰ to +23‰, with lower d-excess occurring during July-October 2010, and the most positive d-excess occurring October 2011. The tap water (**Figure 7**) displays a seasonal pattern similar to the reservoirs (**Figure 5**). Tap water was isotopically the most positive in September 2010 and August 2011, and the most negative from January 2011 to mid-March 2011. Pearson positive correlations of significance (p < 0.05) of the tap water variables existed between temperature-δ^{18}O, temperature-δD, δ^{18}O-δD, and δD-d-excess. A significantly negative relationship occurred between δ^{18}O-d-excess. Tap waters data fall both above and below the GMWL (δD = 6.6*δ^{18}O + 4.1; r^2 = 0.69; **Figure 8**), plotting in a similar manner as the reservoir and precipitation data.

4. Discussion

4.1. Isotopic Records of Central Ohio Precipitation

Records of the isotopic composition of Central Ohio precipitation exist for two previous time intervals: January 1966 to December 1971, which are monthly samples from Coshocton, Ohio [18] and October 1992 to mid-December 1994 for all precipitation events from Oxford, Ohio [19]. These Ohio precipitation data collected in 1966-1971 have a range of δ^{18}O from −17.7‰ to −1.2‰, δD ranges from −126‰ to +3.6‰, and d-excess ranges from +0.64‰ to +24.6‰ and the precipitation data collected from 1992-1994 has a range of δ^{18}O = −20.1‰ to

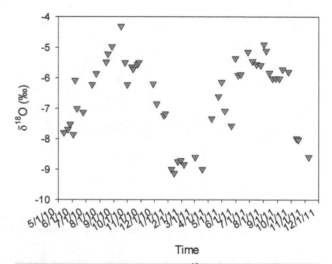

Figure 7. Residential tap waters δ^{18}O signatures distributed from DRWP over May 2010 to November 2011.

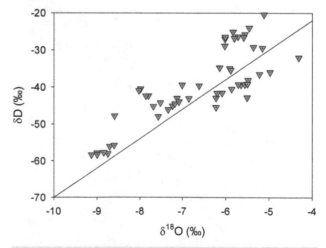

Figure 8. Residential tap water isotopic values from May 2010 to November 2011 plotted in relation to the GMWL.

$-0.47‰$; $\delta D = -147‰$ to $-3.2‰$; d-excess $= -8.6‰$ to $+25.7‰$. These previous Ohio precipitation data collected in 1966-1971 and 1992-1994 also have similar local meteoric water line (LMWL) equations $- \delta D = 7.5*\delta^{18}O - 8.8$ ($r^2 = 0.97$; [18]) and $\delta D = 7.8*\delta^{18}O - 11.2$ ($r^2 = 0.97$; [19]). Compared to our 2010-2011 precipitation equation of $\delta D = 8.0*\delta^{18}O + 15.6$ ($r^2 = 0.96$), our data have a similar slope but a positive intercept. This positive intercept signifies enhanced moisture cycling as the precipitation data plot mostly above the GMWL, while a negative intercept would imply greater evaporation loss, plotting below the GMWL [20] (**Figure 8**).

The monthly $\delta^{18}O$ and δD values of IAEA [18] and Coplen and Huang [19] are similar to ours as well. All data sets follow a similar trend of more negative isotopic values from January to March, increasingly more positive values until June, and then declining to more negative values until December (**Figure 9(a)**). Overall, the monthly mean $\delta^{18}O$ signatures range over ~15‰, the δD over ~120‰, and there is a smaller variation from the mean in the summer precipitation (April-September) than in the winter precipitation (October to March) (**Figure 9(b)**). Colder temperatures in the winter months produce greater amounts of snow with more negative isotopic values, compared to warmer and isotopically heavier rain (Columbus 2010-2011 snow $\delta^{18}O = -24.2‰$ to $-9.4‰$ and $\delta D = -185‰$ to $-58‰$), and this could contribute to the greater isotopic variation across the seasons (**Figure 9**).

These similar trends suggest that the sources of moisture to Central Ohio and the processes affecting isotopic variation in Ohio precipitation have not changed dramatically over the last 45 years. However, the data on an event, and even seasonal basis, can be used, in part, to detect differences in moisture sources throughout the calendar year. The d-excess data from 2010-2011 provide the best information of changes in moisture sources, as these data have a deviation from the expected modern worldwide sample value of +10‰ (**Figure 9(c)**). The mean d-excess of precipitation varies globally with the majority falling in the +10‰ to +15‰ range, signifying meteoric origin (+10‰). In the United States, d-excess cannot be explained by a single parameter, but varies strongly with geographic location. It is very much related to different source air masses, differences in temperature, aridity, and the contribution from evapoconcentration and other moisture sources [21]. Higher d-excess can be caused by evaporation that leaves the remaining water vapor more positive in δD and $\delta^{18}O$ values, with a GMWL slope less than 8 [21]. Precipitation with a large d-excess value may result from the downwind admixture of the evaporated moisture [22]. Therefore, an increased d-excess in precipitation (>+15‰) could be linked to the significant addition of re-evaporated moisture within continental locations. If moisture from precipitation with an average d-excess of 10‰ is re-evaporated, the lighter $^{2}H^{1}H^{16}O$ molecule may contribute preferentially to the isotopic composition of the water vapor and this, in turn, leads to a more positive d-excess in precipitation.

The monthly d-excess values in the Northern Hemisphere follow a U-shaped trend with January d-excess of ~+12‰, steadily dropping to ~+6.5‰ in June, and rising back to ~+12‰ in November [20]. Hence, the d-excess values of >+15‰ during January-April and September-December could signify that the resultant precipitation is from increased moisture cycling from a more inland location, while in May-August d-excess of >+10‰ could indicate that this recycled moisture component is less. If re-evaporation and moisture cycling are the controls of these d-excess values, the possible moisture source could be from the Great Lakes Region, rather than from Ohio's primary precipitation source, the Gulf of Mexico. D-excess in May-June 2010 (+17.7‰ and +19.4‰) and August-November 2011 (+20.7‰, +22.4‰, +20.4‰, +22‰) display consistently higher than +10‰ d-excess values (**Figure 9(c)**). In 2010, the d-excess values represent averages from 11 and 10 sampling events in May and June respectively, due to the high number of precipitation events that occurred in this interval. Water recycling might have played a more important role that resulted in higher d-excess values during this time. In 2011, the higher d-excess values occurred during the wettest fall on record for Columbus with 26 rain events sampled between August to November. This interpretation supports the dependence of the isotopic signatures on the amount effect with the moisture source influenced by more continental sources, as signified by the higher d-excess values measured.

4.2. Comparison to Previous Riverine and Tap Water Studies in Central Ohio

Kendall and Coplen [23] conducted an isotopic study to describe the $\delta^{18}O$ and δD in river waters across the United States, and reported the Ohio LMWL derived from river water to be $\delta D = 5.2*\delta^{18}O - 8.2$ ($r^2 = 0.73$; n = 82). Their river sampling locations within Ohio were located along the state borders, and precipitation was collected in a central northeast section of Ohio. The goal of their work and their river isotopic dataset was to serve as a proxy for the isotopic composition of modern precipitation in the USA. The slope of the LMWL for their

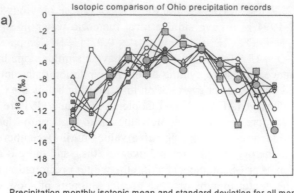

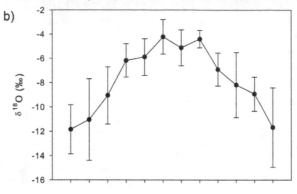

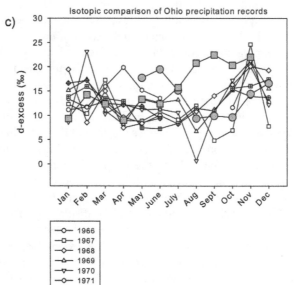

Figure 9. a) Average of monthly $\delta^{18}O$ precipitation signature comparison within Ohio from January 1966 to December 1971 [18], October 1992 to mid-December 1994 [19], and May 2010 to November 2011; b) The Ohio monthly $\delta^{18}O$ precipitation mean with error bars of 1 standard deviation of the mean, using all Ohio data [18] [19]; c) comparison of average monthly d-excess in precipitation within Ohio from January 1966 to December 1971 [18], October 1992 to mid-December 1994 [19], and May 2010 to November 2011.

dataset is lower than our 2010-2011 data (precipitation = 8.0; reservoir = 6.0; tap water = 6.6), reflecting the influence of other processes such as evaporation on the river waters, as well as potentially geographical variations [23]. Our reservoir waters of 2010-2011 have the most similar LMWL slope to their river water. The lower slopes of the LMWLs might imply significant post-rain evaporation of river samples prior to collection; if so, the data would not be representative of local rainfall. A slope lower than 6 in the LMWL for river water is indicative of evaporation, but it is unclear if this occurs during rainfall, within the soil zone as the water moves through the watershed, or in the streams themselves. Kendall and Coplen [23] predict, d-excess values in precipitation of >10‰, whereas the d-excess values of our study range from +1.4‰ to +23‰. The 2010-2011 d-excess has a wide range of values, and the range represents seasonal influences, water vapor sources, and moisture cycling creating higher d-excess. The Kendall and Coplen [23] prediction does overlap with the d-excess range observed in our study, as might be expected for river systems draining larger watersheds. Based on the discharge-weighted means of the spatial isotopic distribution in rivers, Kendall and Coplen [23] predicted precipitation values of $\delta^{18}O$ of −8‰ to −6‰ and δD of −60‰ to −40‰ for Ohio. These compare, but are more negative than our precipitation averages $\delta^{18}O$ of −6.9‰ and δD of −40.2‰. This supports the general contention that large-scale isotopic signatures of precipitation are preserved in the river isotopic compositions in Central Ohio [23].

Bowen *et al.* [1] described the distribution of $\delta^{18}O$ and δD in tap water across the United States as "dominated by spatially patterned variability." Bowen *et al.* [1] demonstrated that spatially coherent patterns in tap water reflect pervasive regionally features of water supply hydrology, and described the tap water isoscape of the United States through isotopic mapping of measured tap waters from across the country. Tap water patterns of $\delta^{18}O$ and δD were explained in terms of water sources and post-precipitation processes (*i.e.* seasonality, recharge, evaporation) affecting surface and groundwater resources.

The predicted isotope composition of Ohio tap water from interpolated precipitation data was $\delta^{18}O$ of −10‰ to −6.1‰ and δD of −59‰ to −49‰ [1]. Tap water samples were collected monthly in Columbus, Ohio, from January 2005 to June 2006 [1] [16]. These data henceforth will be referred to as the 2005-2006 tap waters. These 2005-2006 tap waters ranged between $\delta^{18}O$ of −9.4‰ to −6.1‰, δD of −63.4‰ to −42.6‰, and with a d-excess of +3‰ to +15‰. While the May 2010 to November 2011 tap waters ranged between $\delta^{18}O$ of −8.9‰ to −4.9‰, δD of −57.8‰ to −23.5‰, and d-excess of +1.8‰ to +23‰ for the monthly isotopic mean. Overall, the 2010-2011 tap waters were at times isotopically more positive and spanned a larger range of values than the 2005-2006 tap waters. All monthly average temperatures in both 2005-2006 and 2010-2011 were higher than the 25-yr average monthly temperatures except for January 2006. The monthly precipitation amounts were higher than the 25-yr average precipitation amount only 7 out of 17 months during 2005-2006 compared to 11 out of 19 months in 2010-2011, with 2011 being the wettest year on record. The total precipitation amount over the 17 months during 2005-2006 was 138 cm with no data available for June 2006, and the 19 months over 2010-2011 had a total precipitation amount of 175 cm. It should be noted that 2005-2006 tap waters were distributed from the HCWP using water from Hoover Reservoir, while the 2010-2011 tap waters were distributed from the DRWP with water from Griggs/O'Shaughnessy Reservoir.

In a comparison of 2010-2011 and 2005-2006 tap water data, the waters demonstrated different isotopic trends (**Figure 10(a)**). Seasonal transitions were seen in both tap water studies, but they occurred at different times of the year. The 2005-2006 data demonstrate a seasonal "lag" time in the precipitation-reservoir-tap system, as more negative isotopic winter waters were associated with the summer tap waters and more positive summer waters with the winter. The 2010-2011 tap waters match much better with seasonal isotopic precipitation patterns, with a faster, steeper gradient during the seasonal transition compared to the 2005-2006 tap waters. There appears to be a lag time of ~2 - 4 months when comparing the $\delta^{18}O$ of the 2005-2006 tap waters to the 2010-2011 data, depending on the time of year. This difference in lag time between precipitation and tap could be explained by the difference in reservoir type, innate differences in water flow paths in the different watersheds, or longer residence times of water due to the differences in precipitation amounts, temperatures, and vegetative growth between the years.

All else being equal, the lacustrine type flow system of Hoover Reservoir should have a longer water residence time resulting in a greater seasonal lag and a lower slope of the seasonal isotopic transition. This supports the work of Allen [8] who calculated an average residence time of 152 days in the Hoover Reservoir; in comparison to the calculated O'Shaughnessy Reservoir residence time as an average of 26 days. Hoover Reservoir's capacity is an order of magnitude greater than O'Shaughnessy's, and Hoover's water plant provides a larger

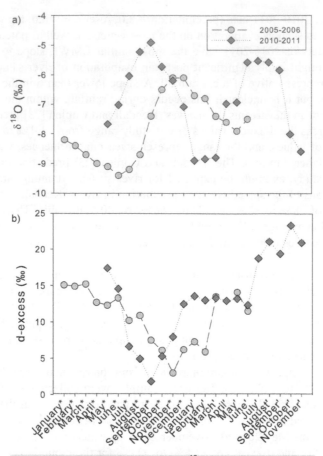

Figure 10. a) DRWP tap water $\delta^{18}O$ signature from May 2010 to November 2011 compared with the January 2005 to June 2006 tap waters distributed from the HCWP [1]; b) DRWP tap water d-excess from May 2010 to November 2011 compared with the January 2005 to June 2006 tap waters distributed from the HCWP [1].

amount of water daily. When comparing each reservoir's capacity to their daily volume of treated water, Hoover's ratio (7.9×10^8 m^3: 4.7×10^5 m^3/d = 16,808 days) is 2 orders of magnitude larger than O'Shaughnessy's (1.8×10^7 m^3: 2.5×10^5 m^3/d = 720 days). These differences support the concept that the O'Shaughnessy/Griggs reservoir system acts more like a riverine system with a smaller capacity but of faster flow-through than the Hoover reservoir system.

In a comparison of the 2005-2006 and 2010-2011 tap waters d-excess values, both waters each follow a V-shaped trend (**Figure 10(b)**). The 2005-2006 tap waters decreased from +15‰ in March 2005 to +2‰ in November 2005, and then increased to ~+13‰ in March 2006 until June 2006. The 2010-2011 tap waters d-excess values decreased from +17‰ in June 2010 to +2‰ in September 2010, and then increased to +14‰ in December 2010 until June 2011. D-excess of tap waters from July 2011 to November 2011 varied over +20‰ to +23‰. The 2010-2011 tap waters have their lowest d-excess in September 2010, while the 2005-2006 lowest d-excess occurs in November 2005. These data suggest a lag time of ~2 months, a shorter lag time than the $\delta^{18}O$ comparison, but both retained the primary precipitation signal. This difference in lag time may be explained by the idea that the d-excess responds more sensitively to seasonal changes in water source, as its main controlling factors are humidity, wind speed, and moisture availability. Also the d-excess seasonal slope transitions were steeper in the 2010-2011 tap waters compared to the 2005-2006 tap waters. This could be a function of each reservoir/water plant system, as the 2005-2006 tap waters came from the lacustrine system (Hoover) with a greater holding capacity and the 2010-2011 waters of the smaller riverine system (Griggs/O'Shaughnessy). The 2010-2011 d-excess tap water values followed closely with the d-excess values of precipitation during May 2011 to

November 2011, with d-excess values >+20‰. This period also coincides with the wettest fall on record, and supports the idea that the Griggs/O'Shaughnessy system retains the precipitation signature even with its "fast" flow-through characteristics, especially during periods of higher precipitation. This isotopic tap water comparison helps describe how differing reservoir types, lacustrine versus riverine, can potentially influence a municipal water cycle.

4.3. Water Cycle of Precipitation to Tap

The precipitation, Griggs reservoir waters, and the tap water are compared in **Figure 11**. The tap water values are similar to the reservoir (**Figure 11**). Even with the large amounts of precipitation during the study period, the reservoir and tap water signatures are similar, with the reservoir level kept at its normal pool level of 258.6 masl. Columbus tap waters have the most negative signature during January 2011 to March 2011, but winter reservoir data were unavailable because of the inability to sample during those times because of an ice cover. However, this lag suggests that the precipitation residence time from reservoir to tap is about two months.

If precipitation is rapidly utilized in the domestic supply and has a short residence time in the reservoir, the $\delta^{18}O$ values should have a similar time series trend. In the comparison of monthly 2010-2011 $\delta^{18}O$ averages of tap waters and volume-weighted precipitation, there are many months (December 2010, February 2011, April 2011, June 2011, and October 2011) during the study period that the tap water $\delta^{18}O$ does not relate well to the precipitation (**Figure 12**). Months that fall on the left side of the tap water: precipitation line have tap waters that

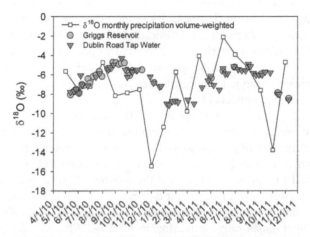

Figure 11. A comparison of Griggs reservoir, tap waters, and precipitation $\delta^{18}O$ volume-weighted signature over May 2010 to November 2011.

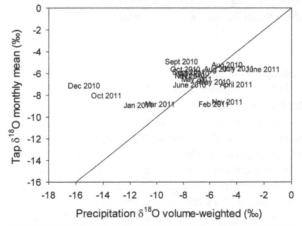

Figure 12. A comparison of tap water $\delta^{18}O$ monthly averages and precipitation $\delta^{18}O$ monthly averages over May 2010 to November 2011.

isotopically resemble summer to fall signatures, while the precipitation was of a more negative isotopic signature, suggestive of winter precipitation. Griggs/O'Shaughnessy Reservoir levels were close to the normal pool levels during both December 2010 and January 2011, while slightly higher in October 2011 at 258.8 masl. This discrepancy suggests some lag from the precipitation falling on the watershed to the tap, possibly due to longer residence times of water in the watershed, reservoir management, or water being drawn from deeper depths, as reservoir had ice covers during December 2010 and January 2011. The more negative values in October 2011 during the wettest fall on record can be explained by the isotopic amount effect, where a greater amount of rainfall produces lower isotopic values with water quickly exiting the reservoir without being captured. The February 2011, April 2011, June 2011, and November 2011 values fall to the right of the 1:1 tap water: precipitation line, with both tap and precipitation reflecting more positive summer isotopic values. These values had precipitation slightly more positive than the tap water, and this suggests mixing with slightly more negative reservoir waters to achieve the tap water signature, as reservoir pool levels were normal. In general however, the 2010-2011 tap waters retain the primary isotopic signal of the precipitation within the riverine system of Griggs/O'Shaughnessy over very short (*i.e.* weekly) time intervals.

4.4. Reservoir Susceptibility to Evaporation

Adeloye *et al.* [3] performed reservoir storage-yield-reliability planning analyses on two multiple reservoir systems, one in England and other in Iran, to investigate the possible effects of reservoir surface net flux from both baseline and climate-change conditions. The behavior of the two systems was different because of the great differences in climate (humid versus semi-arid). Implications for reservoir management and water supply, especially under future climate change scenarios, are important in order to anticipate water resources availability. Through applying different scenarios and using general circulation climate models, it was discerned that reservoirs with lower water yields, designed to meet the seasonal discrepancy between runoff and demand, will be the most prone to climate-change impacts [3].

The very high d-excess in precipitation during 2010-2011, as previously noted, may suggest re-evaporation of the precipitation moisture source to Central Ohio or upwind of Central Ohio during times of higher than average rainfall or at times of higher rates of precipitation. D-excess measurement could potentially be used to describe evaporation effects within reservoirs [24]. In this study, the d-excess values in the reservoirs displayed lower values (<+10‰) in August to December 2010, and the months of April 2011 to November 2011 all had values >+10‰. The summer lowering of the reservoir d-excess did not occur to the same extent in 2010 as in 2011, likely the result of higher than average precipitation amounts. The 2010 reservoir d-excess was the lowest (+2‰) in October 2010, with other times only decreasing to ~+10‰. Also the Griggs/O'Shaughnessy experienced greater d-excess changes than Alum/Hoover, again supporting the idea that hydrologic dynamics of riverine-type compared to lacustrine-type reservoirs can be different and can be seen using isotopic measurement. This suggests that on-going measurements of d-excess in the reservoir could provide information pertaining to both water source and/or reservoir evaporation. Clearly, a better understanding of the impact of precipitation sources on this value is needed before it can be used as a diagnostic tool.

5. Conclusion

This work establishes an isotopic characterization of precipitation, surface waters, and residential tap waters to describe the flow of water in the human-hydrological system of Columbus, Ohio. Signatures of $\delta^{18}O$ and δD in precipitation follow similar trends of previous data from 1970s and 1990s with seasonal characteristics of more negative values during winter months and more positive values during summer months. It was evident that tap waters were not experiencing as large of a seasonal lag as previously observed in tap waters from 2005-2006. The 2010-2011 residence time of precipitation to tap was about two months compared to four months in 2005-2006. While 2011 proved to be the wettest year on record for Columbus, Ohio, one implication of this was increased moisture source cycling resulting in higher d-excess (>+15‰) during April 2011 to November 2011. The higher precipitation rates during 2011 contributed to a shorter residence time of reservoir water to the residential tap, with reservoir morphology also playing a significant role concerning water cycling. Each reservoir system dynamic was important in its water and solute cycling as Griggs and O'Shaughnessy reservoirs act as a riverine system with a faster flow-through while Alum and Hoover reservoirs are more of a lacustrine environment of more constant concentrations and greater mixing. Even though much of this study was conducted during a time with above average precipitation amounts, the basis of understanding about the municipal system of Columbus, Ohio, concerning the travel of precipitation to the reservoir and distribution to a residence has grown. This work provides local water resource managers with information about Ohio precipitation sourcing and a basis of isotopic reservoir dynamics that could aid in the protection of future water resources.

Acknowledgements

Many thanks to Julie Codispoti, J. D. Stucker, and Kelly Deuerling who all assisted with sample collections. We also thank Sue Welch and Chris Gardner for their assistance in the lab. A special thanks to Dave Lape who built the precipitation collector. The School of Earth Sciences, Ohio State University, provided partial support for this work as a teaching assistantship to Deborah Leslie.

References

[1] Bowen, G.J., Ehleringer, J.R., Chesson, L.A., Stange, E. and Cerling, T.E. (2007) Stable Isotope Ratios of Tap Water in the Contiguous United States. *Water Resources Research*, **43**, 1-12. http://dx.doi.org/10.1029/2006WR005186

[2] Hutson, S.S., Barber, N.L., Kenny, J.F., Linsey, K.S., Lumia, D.S. and Maupin, M.A. (2004) Estimated Use of Water in the United States in 2000. USGS Circular 1268. Geological Survey, Denver.

[3] Adeloye, A.J., Nawaz, N.R. and Montaseri, M. (1999) Climate Change Water Resources Planning Impacts Incorporating Reservoir Surface Net Evaporation Fluxes: A Case Study. *Water Resources Development*, **15**, 561-581. http://dx.doi.org/10.1080/07900629948763

[4] Montaseri, M. and Adeloye, A.J. (2004) A Graphical Rule for Volumetric Evaporation Loss Correction in Reservoir Capacity-Yield-Performance Planning in Urmia Region, Iran. *Water Resources Management*, **18**, 55-74. http://dx.doi.org/10.1023/B:WARM.0000015389.70013.e4

[5] United States Census Bureau (2010) 2010 Population Estimates. United States Census Bureau. http://factfinder.census.gov

[6] City of Columbus (2005) Department of Public Utilities 2005 Annual Report. City Council, Columbus.

[7] Wikipedia (2010) Alum Creek Lake. 25 June 2010, 21:27 UTC. Wikimedia Foundation Inc. Encyclopedia On-Line. http://en.wikipedia.org/wiki/Alum_Creek_Lake

[8] Allen, G. (2011) An Analysis of the Fate and Transport of Nutrients in the Upper and Lower Scioto Watersheds of Ohio. Ph.D. Dissertation, The Ohio State University, Columbus.

[9] Lis, G., Wassenaar, L.I. and Hendry, M.J. (2008) High-Precision Laser Spectroscopy D/H and $^{18}O/^{16}O$ Measurements of Microliter Natural Water Samples. *Analytical Chemistry*, **80**, 287-293. http://dx.doi.org/10.1021/ac701716q

[10] Thompson, L.G., Mosley-Thompson, E., Brecher, H., Davis, M., León, B., Les, D., Lin, P.-N., Mashiotta, T. and Mountain, K. (2006) Abrupt Tropical Climate Change: Past and Present. *Proceedings of the National Academy of Sciences of the United States of America*, **103**, 10536-10543. http://dx.doi.org/10.1073/pnas.0603900103

[11] Craig, H. (1961) Isotopic Variations in Meteoric Waters. *Science*, **133**, 1702-1703. http://dx.doi.org/10.1126/science.133.3465.1702

[12] Merlivat, L. and Jouzel, J. (1979) Global Climatic Interpretation of the Deuterium-Oxygen 18 Relationship for Precipitation. *Journal of Geophysical Research*, **84**, 5029-5033. http://dx.doi.org/10.1029/JC084iC08p05029

[13] Jouzel, J., Merlivat, L. and Lorius, C. (1982) Deuterium Excess in an East Antarctic Ice Core Suggests Higher Relative Humidity at the Oceanic Surface during the Last Glacial Maximum. *Nature*, **299**, 688-691. http://dx.doi.org/10.1038/299688a0

[14] Ohio Agricultural Research and Development Center (2013) 1986-2011 Precipitation Records in Franklin County, Ohio, 1986-Present. http://www.oardc.ohio-state.edu/newweather/stationinfo.asp?id=14

[15] National Weather Service/National Oceanic and Atmospheric Administration (2013) 2010-2011 Columbus Ohio Meteorological Data. http://www.wpc.ncep.noaa.gov/noaa/noaa_archive.php?reset=yes

[16] Leslie, D. (2013) The Application of Stable Isotopes $\delta^{11}B$, $\delta^{18}O$, and δD in Hydrological and Geochemical Studies. Ph.D. Dissertation, The Ohio State University, Columbus.

[17] Gat, J.R. and Gonfiantini, R., Eds. (1981) Stable Isotope Hydrology: Deuterium and Oxygen-18 in the Water Cycle. IAEA Technical Report Series #210, Vienna, 337.

[18] IAEA (1992) Statistical Treatment of Data on Environmental Isotopes in Precipitation. Technical Report Series No. 331. International Atomic Energy Agency, Vienna, 781.

[19] Coplen, T.B. and Huang, R. (2000) Stable Hydrogen and Oxygen Isotope Ratios for Selected Cities of the US Geo-

logical Survey's NASQAN and Benchmark Surface-Water Networks. US Geological Survey Open-File Report 00-160; 424. US Geological Survey, Denver. http://water.usgs.gov/pubs/ofr/ofr00-160/pdf/ofr00-160.pdf

[20] Froehlich, K., Kralik, M., Papesch, W., Rank, D., Scheifinger, H. and Stichler, W. (2008) Deuterium Excess in Precipitation in Alpine Regions—Moisture Cycling. *Isotopes in Environmental and Health Studies*, **44**, 61-70. http://dx.doi.org/10.1080/10256010801887208

[21] Sharp, Z. (2007) Principles of Stable Isotope Geochemistry. Pearson Education, New Jersey.

[22] Gat, J.R. (2005) Some Classical Concepts of Isotope Hydrology: Rayleigh Fractionation, Meteoric Water Lines, the Dansgaard Effects (Altitude, Latitude, Distance form Coast and Amount Effects) and d-Excess Parameter. In: Aggarawal, P.K., Gat, J.R. and Froehlich, K.F.O., Eds., *Isotopes in the Water Cycle*: *Past, Present and Future of a Developing Science*, IAEA, Springer, 127-137. http://dx.doi.org/10.1007/1-4020-3023-1_10

[23] Kendall, C. and Coplen, T.B. (2001) Distribution of Oxygen-18 and Deuterium in River Waters across the United States. *Hydrological Processes*, **15**, 1363-1393. http://dx.doi.org/10.1002/hyp.217

[24] Gat, J.R., Bowser, C.J. and Kendall, C. (1994) The Contribution of Evaporation from the Great Lakes to the Continental Atmosphere: Estimate Based on Stable Isotope Data. *Geophysical Research Letters*, **21**, 557-560. http://dx.doi.org/10.1029/94GL00069

Stable Isotopes Studies in the Urucu Oil Province, Amazon Region, Brazil

Eliene Lopes de Souza[1], Paulo Galvão[2]*, Roseli de Almeida[1], Cleane Pinheiro[3], Marcus Baessa[4], Marcio Cabral[1]

[1]Institute of Geosciences, Federal University of Pará (UFPA), Belém, Brazil
[2]Institute of Geosciences, University of São Paulo, São Paulo, Brazil
[3]Institute of Environment and Spatial Planning of the State of Amapá (IMAP), Macapá, Brazil
[4]Leopoldo Américo Miguez de Mello Center for Research and Development (CENPES), Rio de Janeiro, Brazil
Email: *hidropaulo@gmail.com

Abstract

The study area is the Urucu Oil Province, Municipality of Coari, State of Amazonas, Brazil. This research represents a contribution to the hydrogeological knowledge in the northern region of Brazil, particularly in the central part of the Amazon rainforest, where researches on isotopic are still incipient. The primary goal was to determine, by stable isotopes ^{18}O and ^{2}H measurements, interrelationships between surface water and groundwater, in order to understand the origin and mechanisms of groundwater recharge and discharge. For this, samples of rainwater, superficial water and groundwater were collected between June 2008 and May 2009 for stable isotopic analyzes. This understanding is important in cases of eventual contaminations of the area, which could degrade the water resources. The results show that the superficial waters are typically light waters and have meteoric origin, and the groundwater recharge is by direct rainfall infiltration with primary evaporation before reaching the groundwater table in the Içá-Solimões Aquifer System. The isotopic signatures similarities between groundwater and superficial waters indicate both waters' contributions in the streams and, therefore, in the Urucu river.

Keywords

Stable Isotopes, Hydrogeology, Water Management, Surface Water, Groundwater

1. Introduction

The use of stable isotopes in hydrogeology started in the 50s, with the pioneering works of Urey *et al.* and Eps-

*Corresponding author.

tein and Mayeda, in the reference [1]. Through time, these studies have shown to be an effective way for investigating the complex hydrologic system on a range of spatial and temporal scales, providing quantitative information about surface-ground water interactions.

Physically, the stable isotopes are atoms that take the same position in the table of elements, but have a different number of neutrons and, therefore, mass. The most relevant isotopes for atmospheric and hydrologic sciences are ^{18}O for oxygen (corresponding to the most abundant isotope ^{16}O), and ^{2}H (or Deuterium, D) for hydrogen (corresponding to the most abundant isotope ^{1}H) [2] [3]. These isotopes, during evaporation and condensation phases, become enriched in one phase and depleted in the other. This separation of isotopes is named *isotopic fractionation* [3], and quantifying these processes is possible by calculating the large variability of their isotopic ratios ($^{2}H/H$ and $^{18}O/^{16}O$).

Measurements of stable isotopes also require a common standard. For atmospheric applications, the usual standard is the Vienna Standard Mean Ocean Water (V-SMOW), published and distributed regularly by the International Atomic Energy Agency (IAEA). The delta (δ) notation is used to quantify stable isotope as relative ratios and the isotopes values are reported in per mil (‰) to make it easy to compare significant results. Different results from the standard V-SMOW provide information about recharge and discharge processes; flow, mixing and interconnections among aquifers; evaporation; marine influence; and the sources and mechanism of pollution [1].

Reference [4] observed that the $\delta^{18}O$ and $\delta^{2}H$ values of precipitation that have not been evaporated are linearly related by the equation $\delta^{2}H = 8\ \delta^{18}O + 10$. This equation, known as the Global Meteoric Water Line (GMWL), is based on precipitation data from locations around the globe, and has an $r^2 > 0.95$. This high correlation coefficient reflects the fact that the oxygen and hydrogen stable isotopes in water molecules are intimately associated. The slope and intercept of any Local Meteoric Water Line (LMWL), which is the line derived from precipitation collected from a single site or set of "local" sites, can be significantly different from the GMWL.

Many researchers use oxygen and hydrogen isotopes for investigating different situations, such as storm hydrographs [5]-[8], surface-ground water interactions [9]-[11], and integration of upstream waters [12] [13]. These studies demonstrated the usefulness and applicability of stable isotopes as conservative tracers for hydrogeological studies. In Brazil, the first hydrogeological studies using stable isotopes started in the late 60s and early 70s in the Northeast and Southeast regions, and in the Amazon region [14]. In the Northeast region, the studies were developed to know the origin and mechanisms of groundwater recharge, salinization, transit time and age of waters [15]-[24]. In the Southeast region, studies were initiated in the 80s, especially in the state of São Paulo, where the use of groundwater supply stimulated the isotopic characterization in large aquifers, such as the Guarani Aquifer [25]-[27], or the impact of the sealing surface in aquifer recharging [28]. Studies developed in the Amazon region had the objective of calculating the stream flow of the Negro and Solimões rivers and estimating their contributions to the Amazon River [29]. In parallel, the characterization of the isotopic composition of precipitation was analyzed to develop the local meteoric water line. In this context, several isotopic studies were developed [30]-[34], concluding that the rain is conditioned in part by processes of evapotranspiration [32] [33] [35] [36]. Other studies about hydrological cycle and surface-ground water interactions were carried out [30] [37]-[39].

In regards to this paper, the primary goal was to determine, by stable isotopes ^{18}O and ^{2}H measurements, the interrelationships between surface water and groundwater in the Urucu Oil Province, in the Municipality of Coari, state of Amazonas, Brazil, in order to understand the origin and mechanisms of groundwater recharge and discharge. This understanding is important in cases of eventual contaminations, which could degrade the water resources. For this, samples of rainwater (to perform the Local Meteoric Water Line for Urucu-UMWL), superficial water, and groundwater were collected between June 2008 and May 2009 for stable isotopic analyzes.

2. Site Description

The study area is the Geologist Pedro Leopoldo de Moura Operational Base (BOGPM) [*BOGPM—Base Operacional Geólogo Pedro Leopoldo de Moura*], located in the Urucu Oil Province, Municipality of Coari, state of Amazonas, Brazil, 650 km southwest from Manaus, the state capital (**Figure 1**). This province is known by the greatness of its enterprise, becoming a true enclave in the middle of the Amazon rainforest, which has a daily production of about 100 thousand barrels of oil equivalent (BOE) and 10.36 thousand cubic meters of natural gas per day [40] [41].

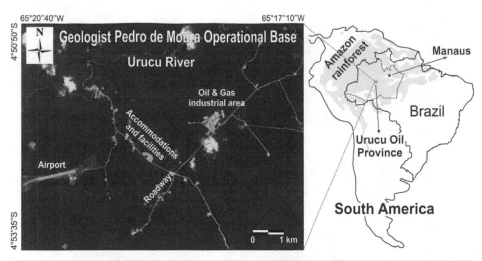

Figure 1. Location map of the study area (modified from Google Earth).

This tropical region is characterized by high precipitation (ranging between 2250 and 2750 mm/year), in which the winter season usually begins in October, with the highest levels of precipitation in January, February and March. The annual mean temperature is 25°C, with a mean seasonal fluctuation of about 1°C. The relative humidity is quite high, between 85% - 90% [41]. Topographically, the region does not have much variation in the elevation, varying between 70 and 90 meters above sea level [41].

The study area is located in the Urucu River Watershed. The main river is the Urucu River, a tributary of the Solimões River, which flows into the Coari Lake. In the BOGPM, there are some creeks, such as Tartaruga and Onça creeks, that flow into the Urucu River [41].

Geologically, the study area is located in the Solimões Paleozoic sedimentary basin [42] [43]. There were five sedimentary depositional sequences on this basin: Ordovician, Silurian-Devonian, Devonian-Carboniferous, Carboniferous-Permian, Cretaceous, and Tertiary-Quaternary [44]. The study area is in the sediments from Cretaceous and Tertiary-Quaternary sequences, represented by the following formations, from bottom to top: 1) Alter do Chão Formation: coarse grained friable sandstones [45] [46]. The age of the upper part of this formation is estimated to be Neo-Cretaceous, by correlation with the Amazonas Basin [42]. The depositional environment is continental, with plain facies and alluvial fans; 2) Solimões Formation: laminated mudstones, lignites layers, and fine to coarse grained sandstones [42]. This formation is estimated to be Miocene/Pliocene, suggesting a meandering fluvial depositional environment and lakes formed by abandoned channels [47]; and 3) Içá Formation: fine to medium grained sandstones and siltstones, with occasional occurrences of conglomerates [48].

Hydrogeologically, the Solimões and Içá Formations, which can have 100 - 120 m thickness, constitute a good water reservoir. The intercalations between these sandstones with clays lenses allow, in some areas, an individualization of two aquifers, hydraulically connected, constituting the semi-confined Içá-Solimões Aquifer System. Right below, the Solimões Aquiclude is constituted by argillite with 150 - 180 m thickness, having the lower contact with the Alter do Chão Aquifer, established by coarse sandstones, good porosity and permeability [49]. The geometry of the Içá-Solimões Aquifer System has a convex shape on the top and a thickness between 50 and 100 m. The hydrodynamic parameters estimated were: transmissivity: 3×10^{-3} m^2/s; storativity: 5×10^{-4}; and hydraulic conductivity: 1×10^{-4} m/s. The potentiometric surface has general flow direction to NNW-SSE, converging to the Urucu River. A considerable cone of depression in the center of the BOGPM is common, due changes in groundwater flow induced by high rate pumping wells [49].

3. Materials and Methods

Water samples were collected during June 2008 and May 2009 in 5 supply wells, rainwater, and 2 surface water points. The location of the sampling points is shown in the **Figure 2**, while the results can be seen in the **Table 1**. The main concern with these samples was to avoid the post-sampling fractionation. For this, pre-cleaned polypropylene vials (30 ml) were used. These vials were completely filled with samples, avoiding air bubbles inside, and stored in coolers, maintaining the temperature, according to the reference [27]. All these samples were analyzed

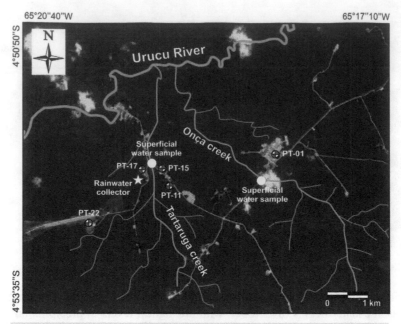

Figure 2. Location map of sampling points, showing locations of supply wells, rainwater collector and surface water at Onça creek and Tartaruga creek (modified from Google Earth).

Table 1. Monthly isotopic concentrations of rainwater, surface water (Onça and Tartaruga creeks) and groundwater samples (supply wells PT-01, 11, 15, 17 and 22), from June 2008 to May 2009.

Water type	Sample location	Stable isotopes (‰)	2008							2009				
			Jun	Jul	Aug	Sep	Oct	Nov	Dec	Jan	Feb	Mar	Apr	May
Rain	BOGPM	$\delta^{18}O$	−6.7	−3.5	−2.2	1.7	-	0.3	−8.5	-	−9.4	−11.9	−13.8	−9.6
		δ^2H	−33.8	−12.6	0.1	19.8	-	13.4	−52.8	-	−57.4	−80.2	−94.2	−63.4
Superficial water	Tartaruga creek	$\delta^{18}O$	−7.0	−4.9	−3.2	−4.2	-	−3.9	−5.6	-	−6.4	−7.8	−8.9	−6.9
		δ^2H	−32.0	−28.1	−15.0	−22.8	-	−18.7	−28.9	-	−37.1	−50.8	−59.2	−43.4
	Onça creek	$\delta^{18}O$	−6.4	−5.5	−4.6	−4.4	-	−4.4	−5.3	-	−6.1	−8.6	−9.9	−6.9
		δ^2H	−31.9	−28.9	−20.9	−22.2	-	−21.1	−28.6	-	−36.7	−56.5	−64.3	−40.8
Grounwater	PT-01 (91 m[*])	$\delta^{18}O$	-	−5.8	−5.5	−5.5	-	-	−5.5	-	−5.7	−5.6	−5.9	−5.9
		δ^2H	-	−32.9	−31.7	−31.5	-	-	−31.1	-	−31.1	−31.4	−33.3	−34.8
	PT-11 (110 m[*])	$\delta^{18}O$	−5.9	−5.6	−5.5	−5.5	-	−5.9	−5.8	-	−4.1	−4.1	−4.9	−4.9
		δ^2H	−28.8	−30.6	−30.5	−30.1	-	−32.0	−32.5	-	−24.1	−24.1	−27.6	−27.3
	PT-15 (120 m[*])	$\delta^{18}O$	−5.6	−4.9	−5.1	−4.7	-	−5.1	−4.7	-	−5.1	−4.8	−5.5	−5.8
		δ^2H	−32.3	−28.7	−30.9	−27.3	-	−28.0	−28.1	-	−31.5	−30.0	−29.4	−32.9
	PT-17 (40 m[*])	$\delta^{18}O$	−5.5	−5.2	−5.1	−4.5	-	−4.3	−5.7	-	−5.8	−5.8	−6.1	−6.1
		δ^2H	−32.0	−30.4	−29.4	−28.0	-	−28.5	−32.2	-	−31.4	−32.6	−32.6	−32.6
	PT-22 (40 m[*])	$\delta^{18}O$	−4.8	−4.5	−5.1	−4.9	-	−5.2	−5.3	-	-	−4.9	-	−4.6
		δ^2H	−26.2	−25.1	−26.6	−26.0	-	−27.3	−27.3	-	-	−27.4	-	−26.2

[*]Depth of the well screen section; - = No sample due technical problem or well maintenance.

for [18]O and ^{2}H at the Center for Nuclear Energy in Agriculture, University of São Paulo (CENA-USP), City of Piracicaba, São Paulo.

Cumulative monthly rain isotope samples were collected to perform the Urucu Meteoric Water Line (UMWL) for a hydrological year. For this, two polypropylene bottles (5 L) interconnected to each other were used and placed in a cooler to keep the temperature and prevent sunlight penetration. The first bottle had a funnel, to catch the rainwater. When the first bottle was completely filled, the rainwater began to flow into the second one through the connection between them. A silicone hose was also attached on the bottle to balance the pressure and prevent exchanges with atmospheric air [28] (**Figure 3**). This method followed the GNIP (Global Network of Isotopes in Precipitation) instruction, where evaporation and loss of lighter isotopes are negligible [50].

The surface water samples were collected in two creeks: the Onça creek and the Tartaruga creek (**Figure 2**). For this, samples were taken from a bridge, sampling waters from the center of them.

The groundwater samples were taken in 5 wells (PT-01, 11, 15, 17 and 22) used to supply the Oil Province of Urucu (**Figure 2**), giving preference to those who had only one well screen section, except the PT-01, which had some well screen sections (the depth of screen section of each well can be seen in the **Table 1**). There was no sampling in the months that these wells were in maintenance, due to technical problems (October 2008 and January 2009). The groundwater samples were taken after removing several well volumes of water, using pumps installed in these wells. This was done to purge the aquifer of stagnant water and to acquire fresh aquifer samples for analysis.

4. Result and Discussion

4.1. Rainwater

In the Urucu Oil Province, the rainy season occurs from December to May, with mean rainfall of 1521 mm, corresponding to 65.3% of annual precipitation. The period with less precipitation occurs from June to September, with 806 mm, equivalent to 34.7% of the annual total. The average annual temperature is 25.8°C, which June has the lowest value (25.3°C), and September the highest (26.3°C). The annual variation is only 1°C, which is common in tropical regions, where changes in air temperature are minimal [51] (**Figure 4**). A comparison

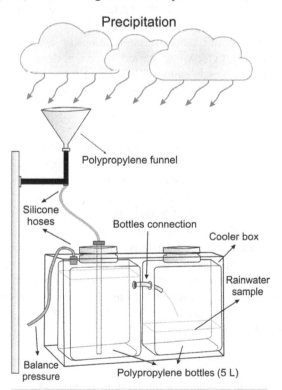

Figure 3. Rainwater collector to perform the local meteoric water line.

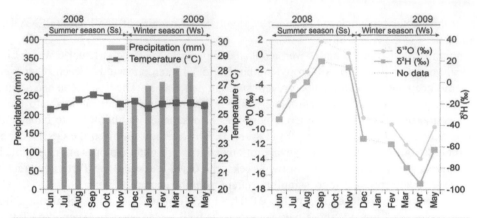

Figure 4. Comparison between climatic data and isotopic concentrations of rainwater, showing an inverse relationship between precipitation and isotopic concentrations.

between monthly rainfall and relative deviations (δ^2H and $\delta^{18}O$) observed that in the month such as March 2009 ($\delta^2H = -80.2‰$; $\delta^{18}O = -11.9‰$) and April 2009 ($\delta^2H = -94.2‰$; $\delta^{18}O = -13.8‰$) showed the minor deviations for both ^2H and ^{18}O (**Figure 4** and **Table 1**). These values, indicative of water with low concentrations of heavy isotopes (light water), coincide with the months of greatest amount of rainfall. Otherwise, the highest deviation, which indicate heavier water, was recorded in September 2008 ($\delta^2H = 19.8‰$; $\delta^{18}O = 1.7‰$) and November 2008 ($\delta^2H = 13.4‰$; $\delta^{18}O = 0.3‰$) (**Figure 4** and **Table 1**). This relationship between isotopic composition of rainwater samples and precipitation was expected because there is usually a depletion of heavy isotopes (water becoming lighter) when the rainfall increase [30] [52].

4.2. Superficial Water

The annual isotopic variation of superficial water relative to the Onça creek ranged between $\delta^2H = -20.90‰$ to $-64.30‰$ and $\delta^{18}O = -4.38‰$ to $-9.90‰$, while in the Tartaruga creek it ranged between $\delta^2H = -15.00‰$ to $-59.20‰$ and $\delta^{18}O = -3.21‰$ to $-8.97‰$ (**Figure 5** and **Table 1**).

The superficial water samples showed the same inverse relationship between precipitation and isotopic concentrations seen in **Figure 4** (**Figure 5**). In the winter period, there was depletion in the concentration of δ^2H and $\delta^{18}O$, while in the summer period there was an isotopic enrichment. This can be interpreted by the surface water evaporation, losing lighter isotopes, or reflecting the type of rainwater that is contributing in the surface water.

4.3. Groundwater

The annual isotopic variation of groundwater samples ranged between: PT-01($\delta^2H = -31.10‰$ to $-34.80‰$ and $\delta^{18}O = -5.49‰$ to $-5.93‰$); PT-11 ($\delta^2H = -24.10‰$ to $-32.50‰$ and $\delta^{18}O = -4.15‰$ to $-5.93‰$); PT-15 ($\delta^2H = -27.30‰$ to $-32.90‰$ and $\delta^{18}O = -4.76v$ to $-5.79‰$); PT-17 ($\delta^2H = -28.20‰$ to $-32.60‰$ and $\delta^{18}O = -4.33‰$ to $-6.16‰$); and PT-22 ($\delta^2H = -25.10‰$ to $-27.40‰$ and $\delta^{18}O = -4.55‰$ to $-5.26‰$) (**Figure 6** and **Table 1**).

In general, the groundwater samples collected both in the deep and shallower zones of the Içá-Solimões Aquifer System did not show a significant pattern related with periods of high or low precipitation, as were noted in rainwater and superficial water samples. However, both samples from shallower zones (PT-17-40 m; PT-22-40 m) and deeper zones of the aquifer (PT-01-91 m; PT-15-120 m) showed a small isotopic enrichment in the summer period, reducing concentrations of δ^2H and $\delta^{18}O$ during winter (water becoming lighter). This behavior follows the same trend seen in the rainwater and superficial water, but with a less accentuated variation. The exception was the PT-11, located in deeper zones (110 m), which showed an isotopic depletion in the summer and a significant enrichment in the winter.

4.4. Relationship among Rainwater, Surface Water, and Groundwater

The isotopic values for rainwater, which represent the Urucu Meteoric Water Line (UMWL), were compared with the Global Meteoric Water Line (GMWL), as well as the Meteoric Water Line for Marajó Island [30],

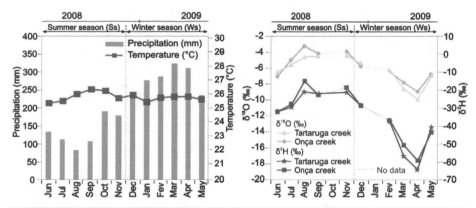

Figure 5. Annual variation of δ^2H and of $\delta18O$ for surface waters in the Onça and Tartatuga creeks, showing an inverse relationship between precipitation and isotopic concentrations.

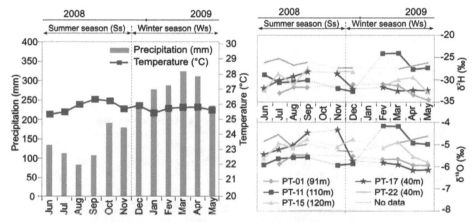

Figure 6. Annual precipitation and isotopic variation of δ^2H and $\delta^{18}O$ for groundwater samples collected both in the deep (PT-01, 11, 15) and shallower zones (PT-17 and 22).

located in the state of Pará, about 1800 km northwest from Urucu, which has similar climate (**Figure 7**). The results showed a similarity between these lines, of which the slope of the UMWL indicated a proximity to GMWL and the Meteoric Water Line for Marajó Island, close to 8. However, the UMWL was a little more negative, indicating waters slightly lighter, in comparison to the others lines, due to Urucu being further from the ocean, causing more isotopic fractionation.

The isotopic concentrations for superficial water showed that the majority of the samples are plotted between the GMWL and UMWL, suggesting a typically light waters and meteoric origin (**Figure 8**), which confirm rainwater contributions in the creeks.

Another comparison was made with waters from the Amazon River, collected in the Marajó Island [30]. According to this study, waters from Amazon River are typically light, due to the most of these waters originating from regions far from the ocean. The comparison showed similarities between both waters, with waters from Urucu slightly lighter than waters from Amazon River. Thus, it is possible to deduce that superficial waters from Urucu have also meteoric origin, with waters coming from regions far from the ocean. The fact that Urucu has lighter rainfall in comparison to the LMWL for Marajó Island (**Figure 7**) also explains lighter superficial waters, in comparison to waters from Amazon River.

The groundwater in Urucu has a different pattern in comparison with the GMWL and UMWL. The majority of the samples were plotted between these lines, with some samples on or near the lines, while others were plotted fairly tightly in the lower left. However, taking into account only the UMWL, the majority of the groundwater samples were grouped below the local line, but close (**Figure 9**). According to the reference [53], this situation shows the possibility of secondary fractionation processes, such as previous evaporation before infiltration or isotopic exchange within the aquifer. So, the facts that all the samples were plotted closer or on the line are likely to be recharged directly from local precipitation with little evaporation.

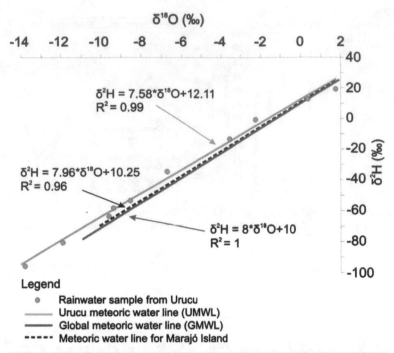

Figure 7. The Urucu Meteoric Water Line (UMWL) in comparison with the Global Meteoric Water Line (GMWL) and the Meteoric Water Line for Marajó Island (Reis *et al.*, 1977). The results have shown a similarity between these lines, with the UMWL little more negative, indicating waters slightly lighter, in comparison to the others lines.

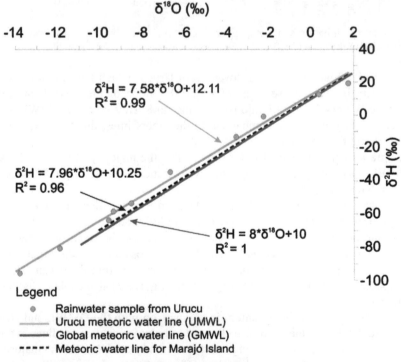

Figure 8. Comparison between isotopic concentrations of surface water from the Onça and Tartaruga creeks with the UMWL, GMWL and waters from the Amazon River, collected in the Marajó Island (Reis *et al.*, 1977). The waters from the creeks are typically light, showing a meteoric origin.

A comparison was made with groundwater samples collected in Marajó Island [30] and in the city of Monte Alegre [39], located in the same region (**Figure 10**). In the city of Monte Alegre, the samples showed similarities with the GMWL, suggesting a current meteoric origin for groundwater. In the case of Marajó Island, samples collected at a depth of 80 m showed heavier isotopic compositions, reflecting different water sources of the aquifer, while samples collected at 5.5 m depth, showed lighter isotopic compositions, similar to the meteoric waters. In Urucu, a significant isotopic variation was not observed between waters from different zones of the aquifer, as seen in Marajó Island, which means that the groundwater in Urucu probably originates from the same water source.

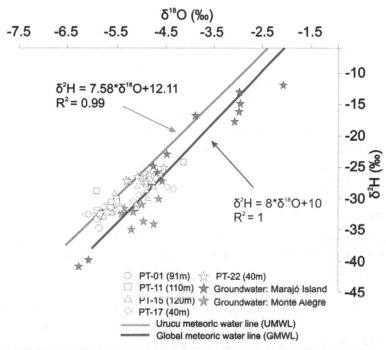

Figure 9. Comparison between groundwater samples collected in Marajó Island [28], in the city of Monte Alegre [39], and Urucu.

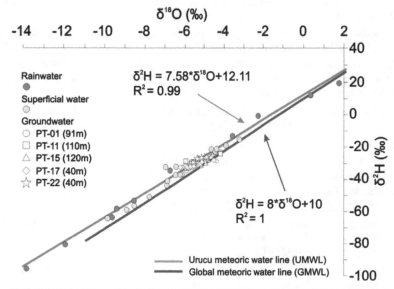

Figure 10. Comparison between rainwater, superficial water and groundwater samples collected in Urucu.

Comparing rainwater, superficial water, and groundwater samples in the same graphic (**Figure 10**), the groundwater recharge is originated from local precipitation, explained by the proximity of the samples in the UMWL. The samples are plotted just below the local line, but very close to it, suggesting previous evaporation of these meteoric waters, depleting isotopically, until reach the Içá-Solimões Aquifer System groundwater table. No difference of isotopic signature between shallow and deeper zones of the aquifer were noted, which means that there is only one water source of recharge in the study area. The isotopic signature of groundwater also coincides with the samples collected in the creeks, indicating groundwater discharging contribution in the streams and, therefore, in the Urucu river.

5. Conclusion

The isotopic values for rainwater in the study area indicated a proximity to the GMWL and the Meteoric Water Line for Marajó Island, but more negative, suggesting waters slightly lighter, due to the location of the study area being further from the ocean, causing more isotopic fractionation. The superficial waters are typically light waters and have meteoric origin, which confirms rainwater contributions in the creeks. The study also suggests that the groundwater recharge is by direct rainfall infiltration. However, the recharging water might have undergone some primary evaporation in the atmosphere or in the soil zone before reaching the groundwater table in the Içá-Solimões Aquifer System. No difference of isotopic signatures between shallow and deeper zones of the aquifer was observed, suggesting one source of recharge in this aquifer, from rainfall. The similarities of isotopic signatures between groundwater and superficial waters were indicative both waters' contribution in the streams and, therefore, in the Urucu river.

Acknowledgements

This work was funded by Petrobras and by Federal University of Pará and Federal University of Pernambuco. A special thanks to geologist Kaitlyn Sunshine Beard for the grammar revision and advises.

References

[1]   Clark, I and Fritz, P. (1997) Environmental Isotopes in Hydrogeology. CRC Press, New York, 328 p.

[2]   Gat, J.R. (1996) Oxygen and Hydrogen isotopes in the Hydrologic Cycle. *Annual Review of Earth and Planetary Sciences*, **24**, 225-262. http://dx.doi.org/10.1146/annurev.earth.24.1.225

[3]   Mook, W.M.E. (2001) Environmental Isotopes in the Hydrological Cycle. Principles and Applications. UNESCO/ IAEA Series. http://www-naweb.iaea.org/napc/ih/documents/global_cycle/Environmental%20Isotopes%20in%20the%20Hydrologic al%20Cycle%20Vol%201.pdf

[4]   Craig, H. (1961) Standard for Reporting Concentration of Deuterium and Oxygen-18 in Natural Waters. *Science*, **133**, 1702-1703. http://dx.doi.org/10.1126/science.133.3465.1702

[5]   Kennedy, V.C., Kendall, C., Zellweger, G.W., Wyerman, T.A. and Avanzino, R.J. (1986) Determination of the Components of Stormflow Using Water Chemistry and Environmental Isotopes, Mattole River Basin, California. *Journal of Hydrology*, **84**, 107-140. http://dx.doi.org/10.1016/0022-1694(86)90047-8

[6]   Buttle, J.M. (1994) Isotope Hydrograph Separations and Rapid Delivery of Pre-Event Water from Drainage Basins. *Progress in Physical Geography*, **18**, 16-41. http://dx.doi.org/10.1177/030913339401800102

[7]   Harris, D.M., McDonnell, J.J. and Rodhe, A. (1995) Hydrograph Separation Using Continuous Open System Isotope Mixing. *Water Resources Research*, **31**, 157-171. http://dx.doi.org/10.1029/94WR01966

[8]   Machavaram, M.V., Whittemore, D.O., Conrad, M.E. and Miller, N.L. (2006) Precipitation Induced Stream Flow: An Event Based Chemical and Isotopic Study of a Small Stream in the Great Plains Region of the USA. *Journal of Hydrology*, **330**, 470-480. http://dx.doi.org/10.1016/j.jhydrol.2006.04.004

[9]   McKenna, S.A., Ingraham, N.L., Jacobson, R.L. and Cochran, G.F. (1992) A Stable Isotope Study of Bank Storage Mechanisms in the Truckee River Basin. *Journal of Hydrology*, **134**, 203-219. http://dx.doi.org/10.1016/0022-1694(92)90036-U

[10]   O'Driscoll, M.A., DeWalle, D.R., McGuire, K.J. and Gburek, W.J. (2005) Seasonal O-18 Variations and Groundwater Recharge for Three Landscape Types in Central Pennsylvania, USA. *Journal of Hydrology*, **303**, 108-124. http://dx.doi.org/10.1016/j.jhydrol.2004.08.020

[11]   Lee, K.S. and Kim, Y. (2007) Determining the Seasonality of Groundwater Recharge Using Water Isotopes: A Case

Study from the Upper North Han River Basin, Korea. *Environmental Geology*, **52**, 853-859. http://dx.doi.org/10.1007/s00254-006-0527-3

[12] Winston, W.E. and Criss, R.E. (2003) Oxygen Isotope and Geochemical Variations in the Missouri River. *Environmental Geology*, **43**, 546-556.

[13] Cartwright, I., Weaver, T.R., Fulton, S., Nichol, C., Reid, M. and Cheng, X. (2004) Hydrogeochemical and Isotopic Constraints on the Origins of Dryland Salinity, Murray Basin, Victoria, Australia. *Applied Geochemistry*, **19**, 1233-1254. http://dx.doi.org/10.1016/j.apgeochem.2003.12.006

[14] Silveira, C.S. and Silva Junior, G.C. (2002) The Use of Environmental Isotopes in Hydrogeological Studies in Brazil: A Critical Review. Yearbook of the Institute of Geosciences, Vol. 25, UFRJ, Rio de Janeiro.

[15] Gat, J.R., Mazor, E. and Mercado, A. (1968) Potential Applications of Isotopic and Geochemical Techniques to Hydrological Problems of Northeastern Brazil. Report to the Atomic Energy Commission and SUDENE, 28 p.

[16] Ferreira de Melo, F.A., Rebouças, A.C., Gat, J.R. and Mazor, E. (1969) Preliminary Isotope Survey of Water Sources in Northeastern Brazil. SUDENE, Spec. Report 18.

[17] Campos, M.M. (1971) Preliminary Survey of Levels of Tritium in Waters of Northeastern Brazil. Report, IPR, Belo Horizonte, 10 p.

[18] Prado, E.B. and Bedmar, A.P. (1976) Contribution of Various Isotopic Techniques to Hydrogeological Study in the Maranhão Basin (Brasil). *Hidrologia Journal*, April-July, 65-79.

[19] Salati, E., Leal, J.M. and Campos, M.M. (1979) Environmental Isotopes Applied to a Hydrogeological Study of Northeast Brazil. SUDENE, No. 58, Recife, 55 p.

[20] Frischkorn, H., Santiago, M.F. and Serjo, A.N. (1989) Isotope Study of Wells in Crystalline Rock of the Semi-Arid Northeast of Brazil. In: *Regional Seminar for Latin America on the Use of Isotope Techniques in Hydrology*, Abstracts, OIEA, Mexico City, 73-89.

[21] Santiago, M.F., Frischkorn, H. and Serejo, A.N. (1990) Isotopic Study of the Waters of Cariri. In: *4th Brazilian Groundwater Congress*, Brazilian Association of Groundwater, Porto Alegre, 338-342.

[22] Silva, C.M.S.V., Santiago, M.F., Frischkorn, H. and Mendes Filho, J. (1996) Distinction between Alluvial Waters and Deep Waters in the Municipalities of Crato and Juazeiro do Norte-CE. In: *9th Brazilian Groundwater Congress*, Brazilian Association of Groundwater, Salvador, CD-ROM São Paulo.

[23] Batista, J.R.X., Santiago, M.M.F., Frischkorn, H., Mendes Filho, J. and Foster, M. (1998) Environmental Isotopes in Groundwater of Picos-PI. In: *10th Brazilian Groundwater Congress*, Brazilian Association of Groundwater, São Paulo.

[24] Costa Filho, W.D., Santiago, M.M.F., Costa, W.D. and Mendes Filho, J. (1998) Stable Isotopes and the Quality of Groundwater in the Plain of Recife. In: *Brazilian Groundwater Congress*, Brazilian Association of Groundwater, São Paulo.

[25] Silva, R.B.G., Kimmelmann, A.A. and Rebouças, A.C. (1985) Hydrochemical and Isotopic Study of Groundwater from the Aquifer Botucatu-Partial Results for the Northern Region of the Paraná Basin. In: *5th Geology Symposium*, Atas, São Paulo, Vol. 2, 489-502.

[26] Kimmelmann, A.A., Rebouças, A.C. and Santiago, M.M.F. (1988) ^{14}C Dating of the Botucatu Aquifer System in Brazil. In: *13th International Radiocarbon Conference*, Abstracts, Iugoslávia, Dubrovinik, 110.

[27] Kimmelmann, A.A., Silva, E., Rebouças, A.C., Santiago, M.M.F. and Silva, R.B.G. (1989) Isotopic Study of the Botucatu Aquifer System in the Brazilian Portion of the Paraná Basin. In: *Regional Seminar for Latin America on the Use of Isotope Techniques in Hydrology*, Abstracts, OIEA, Ciudad de México, 51-71.

[28] Martins, V.T.S. (2008) Application of Isotopes of Pb, Sr, H and O as Tracers for Recharge and Contamination of Metropolitan Aquifers: An Example from the Upper Tietê Basin (SP). Thesis (PhD in Geology), University of São Paulo, USP, IGc, Institute of Geosciences, 220 p.

[29] Matsui, E., Salati, E., Brinkmann, W.L.F. and Friedman, J. (1972) Flows for the Negro and Solimões Rivers through the ^{18}O Concentration. *Acta Amazonica*, **2**, 31-46.

[30] Reis, C.M., Tancredi, A.C.F.N.S., Matsui, E. and Salati, E. (1977) Characterization of Waters in the Region of Marajó through Concentrations of ^{18}O and D. *Acta Amazonica*, **7**, 209-222.

[31] Dall'Olio, A., Salati, E., Azevedo, C.T. and Matsui, E. (1979) Model of Isotopic Fractionation of Water in the Amazon Basin (First Approximation). *Acta Amazonica*, **9**, 675-687.

[32] Salati, E., Dall'Olio, A., Matsui, E. and Gat, J.R. (1979) Recycling of Water in the Amazon Basin: An Isotopic Study. *Water Resources Research*, **15**, 1250-1258. http://dx.doi.org/10.1029/WR015i005p01250

[33] Matsui, E., Salati, E., Ribeiro, M.N.G., Reis, C.M., Tancredi, A.C.S.N.F. and Gat, J.R. (1983) Precipitation in the Central Amazon Basin: The Isotopic Composition of Rain and Atmospheric Moisture at Belém and Manaus. *Acta Amazonica*, **13**, 307-369.

[34] Gonfiantini, R. (1985) On the Isotopic Composition of Precipitation in Tropical Stations. *Acta Amazonica*, **15**, 121-139.

[35] Leopoldo, P.R., Matsui, E., Salati, E., Franken, W. and Ribeiro, M.N.G. (1982) Isotopic Composition of Rainwater and Water from the Soil in the Amazon Forest Bedrock Type, Region of the City of Manaus. *Acta Amazonica*, **12**, 7-13.

[36] Leopoldo, P.R., Matsui, E., Foloni, L.L. and Salati, E. (1984) Variation of the Values of D and ^{18}O Sheet in Water during the Evaporation Process. *Nuclear Energy and Agriculture*, **6**, 3-18.

[37] Matsui, E., Azevedo, C.T. and Salati, E. (1980) Distribution of Deuterium in Surface and Ground Waters in Brazil. *Nuclear Energy and Agriculture*, **2**, 149-165.

[38] Mortatti, J., Martinelli, L.A., Matsui, E., Victoria, R.L. and Richey, J.E. (1987) Isotopic Variation of Oxygen in the Water of River Solimões/Amazon and Its Main Tributaries. *Energia Nuclear E Agricultura*, **8**, 14-23.

[39] Lopes, E.C.S. (2005) Hydrogeochemical and Geophysical Studies of the Braquidobra Region of Monte Alegre/PA. Dissertation (MSc in Geology), Federal University of Pará, Centre of Geosciences, School of Postgraduate Geology and Geochemistry, Belém, 75 p.

[40] Petrobras. (2010) Urucu Oil Province. http://www2.petrobras.com.br/minisite/urucu/urucu.html

[41] Galvão, P.H.F. (2011) Hydrogeological Characterization of the Aquifers of Solimões Formation in the Geologist Pedro de Moura Operational Base, Urucu Oil Province (AM). Dissertation (MSc in Geology), Federal University of Pernambuco, CTG, Graduate Program in Geosciences, Recife, 148 p.

[42] Caputo, M.V. (1984) Stratigraphy, Tectonics, Paleoclimatology and Paleogeography of Northern Basins of Brasil. Ph.D. Thesis, University of California, Santa Barbara, 586 p.

[43] Eiras, J.F., Becker, C.R., Souza, E.M., Gonzaga, F.G., Silva, J.G.F. and Daniel, L.M.F. (1994) Solimões Basin. *Petrobras Bulletin of Geosciences*, **8**, 17-45.

[44] Caputo, M.V. and Silva, O.B. (1991) Sedimentation and Tectonics in the Solimões Basin. In: Raja Gabaglia, G.P. and Milani, E.J., Eds., *Origin and Evolution of Sedimentary Basins*, Petrobras, Rio de Janeiro, 169-193.

[45] Caputo, M.V., Rodrigues, R. and Vasconcelos, D.N.N. (1971) Lithostratigraphy of the Amazon Basin. Internal Report, Petrobras, Belém.

[46] Caputo, M.V., Rodrigues, R. and Vasconcelos, D.N.N. (1972) Stratigraphic Nomenclature of the Amazon Basin: History and Update. In: 26*th Brazilian Congress of Geology*, Brazilian Society of Geology, Belém, Vol. 3, 35-46.

[47] Cruz, N.M.C. (1987) Quitinozoarious from Solimões Basin, Brazil. Internal Report, Belém Covenant CPRM/Petrobras.

[48] Nogueira, A.C.R., Arai, M., Horbe, A.M., Silveira, R.R. and Silva, J.S. (2003) The Marine Influence in Deposits of the Solimões Formation in the Region of Coari (AM): Registration of the Miocene Transgression in the Western Amazon. In: 8*th Symposium on Geology of the Amazon*, Abstracts, Brazilian Society of Geology, Manaus.

[49] Galvão, P.H.F., Demétrio, J.G.A., Souza, E.L., Pinheiro, C.S.S. and Baessa, M.P.M. (2012) Hydrogeological and Geometric Characterization of the Içá and Solimões Formations in Urucu Area, Amazonas State. *Brazilian Journal of Geology*, **42**, 142-153.

[50] IAEA/WHO (2004) Global Network of Isotopes in Precipitation. The GNIP Database. http://isohis.iaea.org

[51] Leng, M.J. (2006) Isotopes in Palaeoenvironmental Research. Springer Verlag, Berlin, 307 p.

[52] Vuille, M. and Werner, M. (2005) Stable Isotopes in Precipitation Recording South American Summer Monsoon and ENSO Variability: Observations and Model Results. *Climate Dynamics*, **25**, 401-413. http://dx.doi.org/10.1007/s00382-005-0049-9

[53] Mazor, E. (1991) Applied Chemical and Isotopic Groundwater Hydrology. Open University Press, Suffolk, 274 p.

3

Employing Water Demand Management Option for the Improvement of Water Supply and Sanitation in Nigeria

Emma E. Ezenwaji[1], Bede M. Eduputa[2], Joseph E. Ogbuozobe[1]

[1]Department of Geography and Meteorology, Nnamdi Azikiwe University, Awka, Nigeria
[2]Department of Environmental Management, Nnamdi Azikiwe University, Awka, Nigeria
Email: emmaezenwaji@gmail.com

Abstract

The aim of this paper is to assess the importance of Water Demand Management (WDM) strategy to the improvement of water supply and sanitation in Nigeria. Persistent water supply shortages and poor sanitation have since remained important features of the Nigerian urban and rural communities. Most often governmental solution to these problems has been to develop and exploit the available water resources and the level of sanitation for the people. This predominant approach which is also known as augmentation method is supply driven with the primary purpose being how best to meet the perceived water and sanitation demand. One of the major disadvantages of this approach is the huge financial involvement associated with it. Conversely, quite recently water resource managers have begun to direct attention on how consumers can be motivated to regulate the amount and manner in which they use and dispose water to alleviate pressure on freshwater supplies. This new approach is known as water demand management. It is demand driven in that consumers determine their own water need. Employment of WDM by consumers especially in water scarce areas as was discussed in the paper will decrease the amount of water use, thereby limiting unnecessary financial expenditure in exploiting new sources to meet the ever increasing demand.

Keywords

Approach, Assess, Demand, Management, Sustainable

1. Introduction

Nigeria, like most developing countries especially in sub-Saharan Africa is facing serious water crises. All signs indeed show that it is getting worse and will continue to deteriorate if not checked by corrective action. The continuously intensifying scarcity of water resources according to [1] is a crucial problem. The primary reason for the water scarcity being experienced in Nigeria could be traced to growing disparity between the decreasing effective supply and increasing demand for water [2] [3]. The causes of this obvious imbalance between urban demand and water supply manifests in the continuous inability of supply quantities to meet demand. This could be traced to the population growth, higher living standards, increased irrigation, urbanization coupled with the impacts of climate change [4]. Water is, therefore, scarce in all its sectors including residential, industrial, commercial, public institutions, agricultural and a range of others. In Nigeria, apart from agricultural water requirements, residential water needs ranks highest in water demand among the sectors. In this paper, we shall as a result, focus attention on the residential water demand and sanitation. This is on the understanding that water consumers in this sector are most vulnerable to the negative impacts of water scarcity and poor sanitation. This is because people daily eat, wash, bath, drink and engage in other water consuming activities at home and the situation is in contrast with the happenings in other sectors where water consumption is not on daily basis because generally weekends especially Sundays are not included in their normal activities. That residential consumers easily contact water borne diseases such as typhoid, dysentery, cholera, and others is no longer in doubt. In addition, the residential water consumers produce wastewater in course of their water using activities from such areas as household bathrooms, toilets wash hand basins and kitchen sinks which are often not properly disposed and as a result create aesthetic nuisance and breeding ground for mosquitoes and other pests.

The traditional policy which has continued to dominate our action towards confronting water scarcity and poor sanitation depend on utilizing new water resources and constructing more sanitation facilities. The transfer of water from remote sources at times as far as 200 km away and the continued exploitation of new resources around us especially the construction of boreholes has since dominated this supply policy. However, the efficacy of such a supply-side measure is being questioned now since the use of water resources have increased extensively and are reaching an impasse because water supply is by its very nature finite.

Apart from the fast depletion of the finite water resources, the supply augmentation has the added disadvantage of being associated with huge financial costs. To avert this problem water providers and their financial supporters have devised a means of managing the finite resources through supply management which is a water conservation method usually adopted by water supplying authorities, where such authorities take decisions on water management without involving the water consumers. This means that the authorities employ such water supply efficiency measures that are entirely within its own means but outside knowledge of the consumers.

In Nigeria, the various State Water Authorities (SWAs) employ this measure to periodically identify and repair water pipelines, assess the water needs of consumers, design tariff structures, reconstruct some water infrastructure, reduce the unaccounted for water (ufw), sensitize and enlighten consumers on the need to adopt water wise measures, increase the regulatory intervention etc. Quite regrettably, the water supply management option has failed to solve the perennial urban water supply problem in Nigeria because most of the SWAs pay lip services to these important functions, some of them fingering lack of funds as the major problem for their poor operation.

To overcome this obvious lack of commitment by the SWAs to properly manage supply, there has appeared a new policy model that aims at maximizing the benefits of the utilization of water resources and sanitation facilities already in use in order to minimize or even eliminate the need for new water supplies and sanitation facilities [1]. This policy option defined as Water Demand Management is gradually becoming popular especially in developed societies and mainly consists of the following actions:

1) the minimization of losses in storage systems,
2) the reuse of water,
3) the containment of water use to avoid waste, and
4) the efficient use of water and sanitation resources [5].

It is in this realization that this paper attempts to explain the continued urban residential water and sanitation problems in Nigeria and seeks to suggest the urgent need to employ the Water Demand Management option for the improvement of these services.

2. Area of Study

Nigeria with a land area of about 924,000 sq·km is a country of immense water supply potential. Annual rainfall figure according to [6] varies from over 3000 mm in the Niger Delta areas and coastal south to less than 750 mm in the extreme northern parts. Using Thornthwaite's water budgeting procedure, [7] computed the climatic water balance for Nigeria using data from 32 stations. The result shows that out of the average rainfall over the country of 1397 mm, about 1067 mm of this are lost through evapotranspiration process leaving 330 mm for surface and sub-surface run-off. The runoff coefficient of the country as a result is, therefore, about 23.64%. In terms of surface water resources, the country is well drained with a very reasonable network of rivers and streams (**Figure 1**).

Altogether, the surface water resources stock of the country is estimated at 267.3 billion cubic metres [8]. However, it has been observed that the Nigerian rivers are generally turbid because they carry a lot of sediment load in suspection particularly during the rainy season. Some river water in Nigeria is acidic and somehow hard, and heavily biologically and chemically polluted as a result of riparian and watershed activities. It is, therefore, necessary that water from this source should be fully analysed and treated before consumption. Furthermore, Nigeria is rich in ground water resources. Quite often its quality and quantity depends on geology, climate and the underlying rocks. According to [7] boreholes sited in basement complex rocks do not yield as much water as those of sedimentary rocks. For example, yield of wells and boreholes in some locations in the cretaceous sedimentary rocks yield up to 22,730 litres/hr lending support to the findings by [8] that the ground water potential of Nigeria is about 51.9 billion cubic metres.

From the foregoing, it could be seen that Nigeria is well endowed with water resources. The issue then is not that of availability, but that of proper management. For instance, the total average annual surface run-off over the country computed using appropriate water budgeting method is about 330 mm as we had earlier noted and this translates to 835,654 million litres a day. The current residential water needs of the country computed with trend based method shows that with a population of about 160 million (projected from 2006 figure) and based

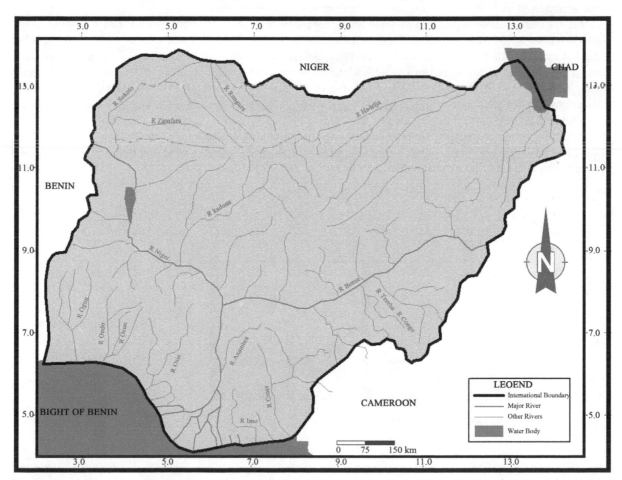

Figure 1. Nigeria's river systems.

on per-capita daily water use of 120 litres (Federal Government recommended level) is then 19,200 million litres/day. This represents 2.22% of the surface water potential of the country of 835,654 million litres/day which includes the underground water resources that is not as exploited as the surface water source. The provision of inadequate toile facilities in Nigeria and the fact that even when there is a flush toilet water to flush after use has increasingly become a problem. Graphs showing cases of water borne diseases associated with lack of potable water supply such as cholera, typhoid fever and dysentery indicate that in 14 selected states studied by the National Bureau of Statistics the situation is to say the least, worrisome.

3. Methodology

Data were collected from secondary source primarily the National Bureau of Statistics (NBS), Nigeria published in 2010. Data which were collected are on cases of water borne diseases in the 36 States of Nigeria and 6 geopolitical zones on the type of toilet mostly used by residents, main source of water supply, quality of water supply, etc. simple graphs such as bar and line charts were utilized in the analysis of data with the purpose of establishing patterns and relationships of identifying water related problems. Furthermore, oral interviews were conducted to verify some areas with doubtful figures in the public data.

4. Results

Figure 2 are graphs showing cases of water borne diseases such as cholera, typhoid fever, in the following selected state where data were available. The States are; Akwa-Ibom, Bauchi, Borno, Cross River, Edo, Ekiti, Gombe, Kwara, Ondo, Osun, Oyo, River and Taraba.

From **Figure 2**, it could be seen that Taraba ranks highest followed by Borno and Gombe States. Conversely the situation in such States as AkwaIbom, Ekiti and Edo are commendable.

Figure 3 indicates the position of some States regarding the use of different types of toilet ranging from very modern ones to the most obsolete. Types of toilets studied regarding their use are water closet (WC), pit latrine, bucket system, toilet facility in another dwelling, public toilet and nearby bush. Among them, pit toilet is the most widely used. The figure shows that Kaduna State has the highest patronage of pit latrine, but generally its usage is highest in all the selected States. The use of pit latrine is followed by that of nearby bush. These are indications of poor sanitation which expectedly gives rise to the prevalence of water borne diseases which we saw in **Figure 2**.

Again the cases of malaria outbreak in the States of the federation including Abuja shows that Imo State tops the table in 2007 with over 600,000 cases, Kastina in 2006 and 2005 had over 380,000 and 270,000 cases respectively, Osun with almost 400,000 and 300,000 cases in 2004 and in 2003, Kano with about 250,000 cases.

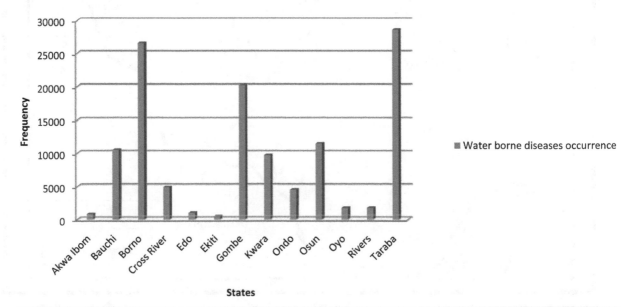

Figure 2. Bar chart of cases of water borne diseases in Nigeria (2008).

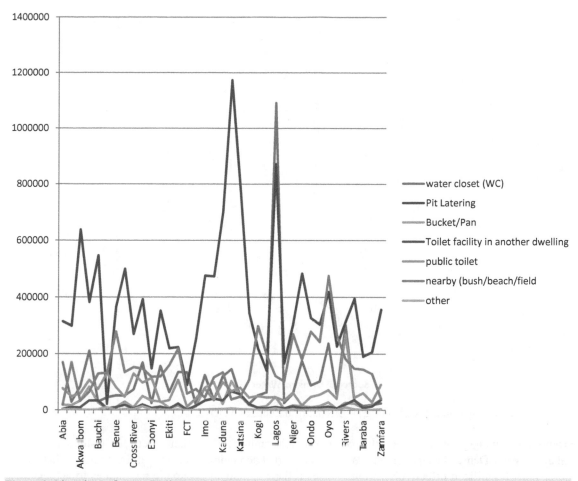

Figure 3. Line chart of type of toilet used in States in Nigeria.

The liquid wastes and other stagnant pools of water that are seen in many locations in our urban areas are ready breeding grounds for mosquitoes. **Figure 4** shows the regional distribution of households by main source of water supply in Nigeria for domestic purpose.

It is seen in the figure that the North East zone ranks highest in the patronage of unprotected wells, followed by North West zone, South East and North Central tie as the highest in the reliance to rivers, lakes or ponds as sources of water. Southwest ranks highest in the use of treated pipe borne source. Generally all the zones are poor in harvesting rainwater for domestic water uses. The situation for individual States that make up each of the regions are shown in **Figures 5(a)-(f)**.

In North East region, the use of pipe borne water is generally low while that of the unprotected wells are patronised heavily by all States in the region. In the Northwest patronage of treated pipe borne source is very poor. All the States in this region patronize unprotected wells with Zamfara State having a slightly higher figure than Sokoto. In North Central region, treated pipe borne water is patronized by all States with FCT having the highest. All the States in this zone have a high patronage of rivers, lakes or ponds with Benue ranking highest. However, in the use of water vendors, FCT stands out, in South East region, boreholes and hand pumps are well patronized with Abia ranking highest. All the States in this region patronize rivers, lakes and ponds. In the South West region, boreholes and hand pumps are equally heavily patronized with Lagos ranking highest. Finally, in the South-South region, boreholes and hand pumps receive the highest patronage followed by rivers, lakes and ponds with Bayelsa ranking highest in the use of rivers.

5. Discussion

The high level of water scarcity and high affliction of consumers with water borne disease have provoked the

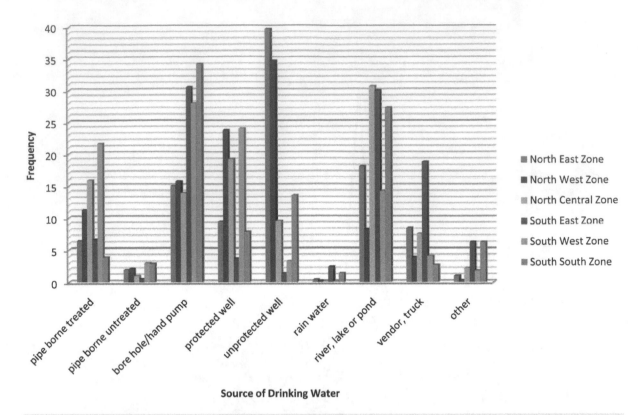

Figure 4. Sources of drinking water in the six Geopolitical Zones of Nigeria.

promotion of management strategy that can reasonably maintain the balance between water supply and demand balance. Water Demand Management (WDM) is aimed at achieving development by stressing that every drop of water must be utilized effectively. It emphasises reliance on socio-economic techniques to achieve the desired objectives. The focus of WDM strategy is more on consumers than on the supply of water from additional water sources.

WDM, therefore, strives to achieve water use efficiency in order to ensure sustainable and affordable water services [9]. The major reasons for WDM according to [10] include:

1) There is a limited environmental capacity of water resources system.
2) There is an ever increasing use of water, whereas the resources are limited such that with increased utilization, supply is diminishing.
3) There are increasing cost of developing new water sources especially as the easily accessible and cheap sources have already been develop. Countries are, therefore, experiencing limited investments due to financial constraints.
4) Water shortages are already occurring worldwide and the existing water development projects cannot be altered to meet the challenges facing many countries. Also [2] added that
5) WDM is usually employed to avert water crisis.
6) It is usually used by Water Corporations to buy time, delaying the need for large scale capital investment for the expansion of water supply. This is particularly of interest when capital for such development is a soft loan or credit from a Development Bank of a bilateral source. In most cases, the savings achieved by delaying an investment can provide financial resources that would more than cover the costs of implementing a comprehensive water demand programme.

The slow progress being made by WDM in the residential water and sanitation is as a result of a range of reasons, some of which are discussed here.

5.1. Lack of Political Will

One of the problems of WDM in solving the urban and rural residential water demand is lack of political will by

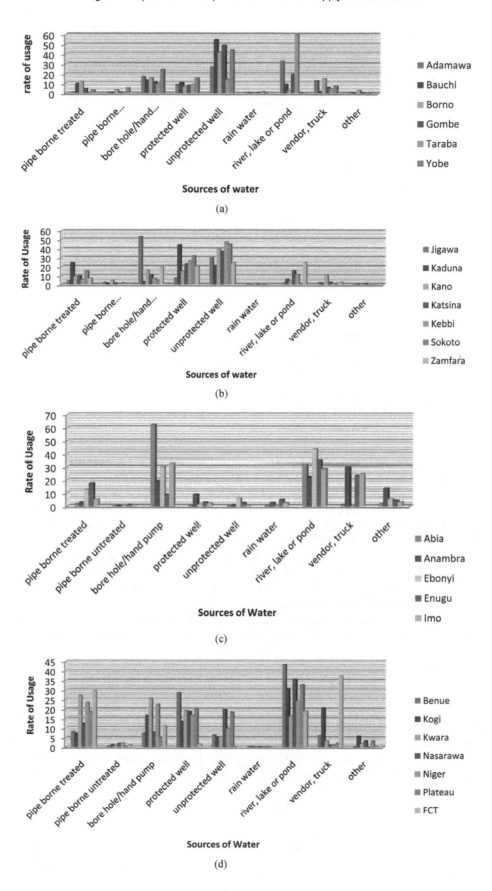

(a)

(b)

(c)

(d)

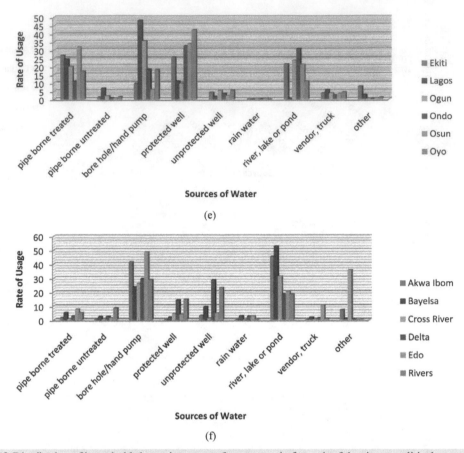

Figures 5. (a)-(f) Distribution of households by main source of water supply for each of the six geopolitical zones of Nigeria.

the politicians who control the State and Local Governments to embark on water management strategy that would benefit the people. They handle the programme whenever it exists with a great measure of unseriousness. Our investigation revealed that the politicians who are at the helm of water supply affairs both in the State, Local Government Areas and various departments are always faced with the dilemma of giving support to consumers on any unpopular water wise measures and that of leaving them to do otherwise. The appointment of unprofessional as commissioners for Public Utilities, a Ministry that oversees the activities of the Water Corporations in various States complicates the problem.

Most of these appointees usually assume their responsibilities with the intension of enriching themselves and any programme of the Water Corporation which is not in line with that aspiration is usually often ignored.

5.2. Lack of Sustained Awareness Campaign

The awareness level of WDM in various States Water Corporation is still low. To buttress this, the government's budgetary proposals have failed to provide the desired importance of this aspect of water management, conveying the impression that the concept is either not well understood or that its effects are not appreciated. It is important to understand that one of the outstanding ways by which WDM could succeed in any society is the ability of the people in that area to change their attitudes to water consumption. Based on this, any water authority or urban government that is unable to do this will not achieve much water conservation under WDM effort.

5.3. Lack of Co-Ordination among Ministries

For WDM option to achieve the desired success, it must be handled holistically. This is unfortunately what is lacking in the States where relevant ministries such as industry, commerce, education, environment, health and technology are not involved in taking the policy decision that would guide water consumers. Since water resources touch on a range of sectors, Ministries that are responsible for the above outlined areas should be involved in the discussion and assessment of issues that would lead to any water demand management policy. Many States Ministry of Public Utility or Water Resources which are supposed to serve as coordinating points have failed in this regard. The various States Water Corporation which are usually mere parastatals of either the Ministry of Public Utilities or Water Resources lack the capacity especially in terms of funding and authority to deal with other Ministries. The result of this is that the WDM lacks proper and closer supervision in most instances.

5.4. Poor In-School Education

In-school education as an instrument of WDM can be effective if well handled. Regrettably in some of our States only few schools undertake this responsibility. In many States, water education has not been made part of the school curriculum. Water Education should be given a good space within the social studies or civics scheme in our schools, with a view to developing attitudinal change on the children. When children are educated on water issues, families are invariably educated, as knowledge gained by these children would be carried back home and implemented.

5.5. Lack of Monitoring Tools for WDM

Many States Water Corporations lack modern monitoring tools to assist them in the monitoring of the WDM especially at the city level. For example, GIS/data base can be used advantageously to present data on flow rates and pressures to any district of any rural community or urban area. This could be used by network managers to detect potential leaks in parts of the city that are experiencing high pressures that could result in leaks and bursts. Furthermore, the Corporation does not have any type of database which could be used to further effects in quantifying leakages especially in terms of revenue.

5.6. High Cost of Retrofitting Technologies

The cost of retrofitting technologies are usually too high in the market and most times these high costs make them fall out of the reach of many residents. Added to this is the cost of labour in installing them. Our field investigation revealed that consumers are sceptical over retrofitting programmes as a result of these cost implications. This problems can be solved if the government can bear some of the material costs *i.e.* offer subsidy to enable people acquire them. Again, the programme should be made to start on a pilot scale, in the States.

WDM as a water management concept has been practised in some residential areas in our urban and rural communities. [11] explained the way it has been practiced in three low income districts of Enugu urban areas namely Ugbene in Abakpa Nike and Agricultural quarters in Asata mine road, Uwani and also in Nkpor district of Onitsha urban area. According to him, residents in these three urban areas organized themselves into Community Based Organizations (CBOs) to monitor compliance of consumers to WDM practices and noted that within a period of two years it was monitored, water demand decreased by 25% in affected districts of Enugu and 35% in the affected district of Onitsha. However, Schouten and Moriarty (2006) were of the view that a key area affecting the management of community water supplies is the resource based itself. How reliable is it? Is quality sufficiently high and if not what can be done? Is there enough water for all domestic activities? Conversely, the issue of water demand is diametrically opposed to supply and their management techniques differ to the same extent. Even within the domain of demand, the residential water demand management seem to be at the root of the solution to the great dividing line between the quantity of water demand and supply. Whenever there is this great demand-supply gap, sanitation within the household suffers considerably. Based on this, we shall examine the water demand management in Nru Nsukka urban area, to show the effects of community approach to Water Demand Management.

The residents of Nru in Nsukka urban area have for long been battling with the best way to provide sufficient water supply and sanitation for its people. According to [12] the ward was able to attain only 22% per capita water supply in 1997 of the Federal Government recommended minimum of 115 litres per capita per day (lcd). Again [13] noted that the 2009 per capita water demand level of the people was only 20.2 lcd, which was about 5 litres less than the situation in 1997. The reason for the above situation is the low level of urban water supply to the area by the Water Corporation. Faced with this problem, members of the community decided to take their destiny in their own hands by trying to develop and manage the available water quantities. Based on this, the community formed an organization in April 2009 which they called the Water Service Association and charged it with the responsibility of ensuring the provision of more water by households and managing same. The Association's first task was on how to improve supplies. On the basis of this thinking, they agreed to embark on rain water harvestation within the community based on the following methods:

1) Collection of rain water from all the zinc roofs of buildings in the community
2) Collection of run-off into shallow ponds or catch pits
3) Collection of rain water running down the stem of certain trees e.g. cocoanut tree.

In doing this, they ensured that all the houses located close together were connected to each other through water catchments made of zinc which are directed into a locally dug tank of high storage capacity. There are about 92 of such high storage tanks of 10,000 litres each.

Water collected in these catch pits and those from the stems of trees are used for less sanitary domestic activities. The entire water collection from zincs between April and October 2010 was estimated at 920,000 litres which gave 40 litres of water per person per day for 24,000 residents of the area which is enough to satisfy hygienic requirements of residents. The water demand management they adopted are outlined as follows:

1) Payment of fine for anybody who do not re use water for the activities where it could be done.
2) Imposition of fine on any consumer who embarks on lawn watering of any kind with water which is adjudged good for hygienic purposes including drinking and cooking.
3) Attracting the government to sensitize their people on the need to build low water or even nil water demand toilets. In this regard, ventilated improved (VIP) latrine has been popularised in the community. The advantage is that this type of toilet uses no water and at same time prevents flies and odour. In this way, the community achieved a low feasible technology option in sanitary requirements especially regarding human wastes.
4) Also the committee was able to develop a local tariff structure which accords with the conventional ones. In this regard, houses with modern water consuming facilities such as flush toilets, bath, shower system and wash hand basins are meant to be connected to the urban water supply to use the small quantities supplied to them by the water Corporation, but will pay 20% above the official government water tariff. This 20% which are normally termed as administrative charge are paid into the community's account and used to develop its water supply system.
5) The community raised a non interest loan from well-meaning members of the community to construct rain catchments and other water supply infrastructure.
6) Others who do not have water consuming facilities in their homes are meant to use water from the community water supply system and pay a flat rate of ₦100.00 per month.
7) The water service committee was also charged with the responsibility of local sanitation. Anyone who does not dump domestic refuse at the officially designated sites are fined. Again defecation in the bushes around peoples homes are forbidden, as offenders are sanctioned. With this, the committee ensures that the community water reservoirs and catch pits are relatively less polluted of faecal and physical matters.
8) Furthermore, the committee appeals to the residents to ensure that their water storage facilities at home are of minimal sanitary standard.
9) General water management discussions are carried out at the monthly town meetings. During this meeting, the committee members and sometimes the Water Corporation's officials are invited to talk on the best water wise uses at homes. This constant interaction with Corporation's officials help to expose every member of the community to water management practices.

The Nru Nsukka Water Demand Management operation serves as an eye opener to the ability of the local people to manage their water and sanitation. From available verifiable results, it was found that within just one year of its operation water demand in the community decreased by 10%, by the second year it has reduced to 19% and the third year demand was down by 25%. This, therefore, means that the community's average per capita water demand of over 140 litres per capita per day (lcd) came down to about 105 lcd just in 3 years. With more of the same efforts in sanitation, it was observed that there was an improved environmental quality as there was less pollution in water supply leading to about 20% reduction of the people afflicted by water borne diseases [14]. This result has indeed shown that it is not only the use of economic instruments that could achieve rational water demand reduction but other measures including sustained community effort. The emphasis on the centrality of water pricing mechanism as a water demand strategy as proposed by many authors including [15]-[17]

focused primarily on the use of price to force down demand. The use of economic instruments for water demand management however, tends to work better in an area with a reliable and uninterrupted water supply as is the case in western cities. Many other studies have actually indicated that it is still difficult to ascertain if market based instruments applied to water demand management achieve the often lofty ambition of decreasing demand. The application of economic instrument in an environment without adequate water supply will surely not succeed as the consumers behaviour tend to ignore any price attached to a good which is not rightly supplied.

Also it has been found from the Nru-Nsukka studies, adequate sanitation could still be achieved in a water scarce environment by the adoption of improved toilet facilities and other less water demanding activities. Also it was observed that efforts at environmental education has the tendency of shifting the attitude of consumers from their original environmental mind-set to a new one in which the people now acquire positive attitudes to deal with the environment.

Community management of water supply is the only thing that can stem the widening water demand-supply gap being experienced by residents in both urban and rural communities in Nigeria. Communities should be allowed to initiate, develop and manage their water supply and sanitation. It is only under this condition that water and sanitation would be rationally managed to the extent that the aim of WDM could be achieved which is the water management that employs various measures to control waste and undue consumption. This can only succeed through community management of water and sanitation which in the South Eastern parts of Nigeria has a ready framework in place. This is because communities in the area are administered through town unions where members of the community are provided with the opportunity of asking questions and contributing to the improvement of any community project including water and sanitation projects. The Nru-Nsukka model has two advantages which are:

1) Government at various levels (Federal, State and Local Governments) should use this approach as a way of reducing government expenditure in community development.

2) Donors will look at it as an opportunity to focus and stretch development budgets towards effective implementation of water supply and sanitation facilities and to bypass the problem posed by corrupt and inefficient government officials [18].

The Nru-Nsukka example is one case of the adoption of the Demand Responsive Approach (DRA) to water and sanitation, a concept which is geared towards putting community management approaches into effect [19] [20].

6. Conclusion

Water Demand Management is a management concept that motivates consumers to regulate the amount and manner in which they access, use and dispose water to alleviate pressure on fresh water supplies. This approach is demand driven as against the old management method which is supply driven. As we have seen, Nigeria spends whooping sums of money to develop new sources from its rich water reserve but these are often misused by the application of poor management strategy. We have realised that a supply side approach in combination with weak and fragmented institutional structure as we have in Nigeria offer no real hope for improved water and sanitation. Community management approach is both demand driven and engages in often pragmatic demand management measures. This equips the community with home grown measures towards addressing the huge unmet water and sanitation needs of the people yawning for these services. Poor water resources management gives rise to poor sanitation and time to employ WDM to improve these services in Nigeria is long overdue.

References

[1]   Bithas, K. and Stoforos, C. (2006) Estimating Urban Residential Water Demand Determinants and Forecasting Water Demand for Athens Metropolitan Area, 2000-2010. *South Eastern Europe Journal of Economics*, **1**, 47-59.

[2]   Ezenwaji, E.E. (2009) Municipal and Industrial Water Demand and Supply in Enugu Urban Area, Nigeria. Unpublished Ph.D. Thesis, University of Nigeria, Nsukka.

[3]   Ezenwaji, E.E. (2012) Water Demand—Supply Gap in the Residential Sector of Enugu Urban Area and Implications for achieving the Millennium Development Goals (MDGs). *The Colloquium on Thirty Years of Social Services in Nigeria*, Benin, 25th-28th March 2012.

[4]   Kayaga, S. and Smout, I. (2007) Water Demand Management: A Key Building Block for Integrated Resource Planning for the City of the Future. *Switch*, **8**, 30-35.

[5]   Renwick, M. and Archibald, S. (1998) Demand Side Management Policies for Residential Water Use: Who Bears the Conservation Burden? *Land Economics*, **74**, 343-359. http://dx.doi.org/10.2307/3147117

[6]   Ayoade, J.O. (1981) On Water Availability and Demand in Nigeria. *Water Supply and Management*, **5**, 361-372.

[7]   Ayoade, J.O. (1975) Water Resources and Their Development in Nigeria. *Hydrological Science Bulletin*, **20**, 581-591. http://dx.doi.org/10.1080/02626667509491589

[8]   Irokalibe, I.J. (2008) Water Management in Federal and Federal Type Countries: Nigerian Perspective. *Forum of Federations, Zaria*, **2**, 49-60.

[9]   Cech, T.V. (2005) Principles of Water Resources: History Development Management and Policy. John Wiley and Sons, New Jersey.

[10] Ezemonye, M.N. (2007) Strategies for Water Demand Management: State of the Art. Doctoral Research Seminar II, Department of Geography, University of Nigeria, Nsukka.

[11] Ezenwaji, E.E. (2003) Urgent Water Demand Management in Nigeria. *Proceedings of the 29th WEDC International Conference*, Abuja, 22-26 September 2003.

[12] Obeta, M.C. (1997) Spatial Patterns of Residential Water Demand and Supply in Nsukka Urban Area of Enugu State, Nigeria. Unpublished Master's Thesis, University of Nigeria, Nsukka.

[13] Anyaorah, B.O. (2010) Water Demand in Selected Wards of Nsukka Urban Area, Nigeria. *Urban Water Affairs*, **16**, 114-122.

[14] Ugwoke, S.N. (2011) Personal Communication.

[15] Espey, M.J. and Shaw, W.D. (1997) Price Elasticity of Residential Demand for Water: A Meta Analysis. *Water Resources Research*, **33**, 1369-1374. http://dx.doi.org/10.1029/97WR00571

[16] Renzetti, S. (2005) Economic Instruments and Canadian Industrial Water Use. *Canadian Water Resources Journal*, **30**, 21-30. http://dx.doi.org/10.4296/cwrj300121

[17] Cantin, B., Shrubsole, D. and Ait-Ouyahia, M. (2005) Using Economic Instruments for Water Demand Management: Introduction. *Canadian Water Resources Journal*, **30**, 1-10. http://dx.doi.org/10.4296/cwrj30011

[18] Schouten, T. and Moriarty, P. (2006) Community Water, Management from Systems to Services in Rural Areas. ITDG Publishing, London.

[19] Sara, J. and Katz, T. (1997) Making Rural Water Supply Sustainable. Report on the Impact of Project Rules, UNDP—Word Bank Water & Sanitation Programme, Washington DC.

[20] World Bank (2002) Rural Water Supply and Sanitation Lacks the Information on Demand Responsiveness (DRA). World Bank Report, Washington DC.

4

Pollution of Wells' Water with Some Elements, Fe, Mn, Zn, Cu, Co, Pb, Cd, and Nickel in Al-Jadriah District, Baghdad Government

Hussein Mahmood Shukri Hussein

Biotechnology Research Center, Al-Nahreen University, Baghdad, Iraq
Email: shukrihussein3@gmail.com

Abstract

Ninety six water samples were collected from eight wells in Al-jadriah district—Baghdad, from June 2010 to May 2011, and analyzed for presence of Fe, Mn, Zn, Cu, Co, Pb, Cd, and Nickel, using Atomic absorption method. Results revealed presence of only scanty amount of iron, Mn and Co, ranged from 0.09 - 0.29 ppm for iron and 0.016 - 0.339 ppm for Mn, and 0.01 - 0.732 ppm for Zn. Concentrations of other elements (Cu, Co, Pb, Cd, and Ni) were nil. All values were below the safe limit of water suitability for human uses, according to safe limits laid down by WHO (2011), Iraqi Measurement and Quality Control System and Iraqi Ministry of Environment. Wells' water can be used under special management in Water management units.

Keywords

Wells' Water, Trace Element, Heavy Element

1. Introduction

Al-Jadriya province is considered high fertile loam soils. It represents the eastern side of the shoulder of the Tigris River and stretching from the eastern section of the province of Baghdad through the park Abu Nawas until the Jadiriya Bridge, and Baghdad University camp to the southern Dora district which is an extension of the plain Iraqi sedimentary [1]. The Tigris River is the only source of irrigation in Jadriya, for homes, small orchards and farmland. However, the reduction in level of Tigris River in recent years pushed those to dig wells to get water for irrigation in the lands and to avoid dependence on the central pipelines network, taking advantage of rich groundwater in the province of Baghdad, especially Jadriya and Karrada district.

Water quality means different things to different people depending on the objectives and purposes of water use. Good quality water should have several specifications and standards as drinking water requires high specifications and standards, and irrigation water requires less specification, while animals breeding (cattle, sheep, poultry, fish), require specifications and standards approaching the drinking water standards for humans.

WHO standards and determinants for drinking water [2] are followed by different countries in the world and each country has its own guide. In Iraq, specifications and standards for drinking water were identified by Central Agency for Standardization and Quality Control and by Ministry of the Environment [3].

It is important to estimate the concentration of trace elements in water to determine the quality of water and their effect on plant growth and production, growth of Microorganisms and human health, but when they exceed safe limits they will be potentially harmful and toxic [4]-[6].

As part of national survey of geological and water quality assessment program in the United States, concentration of Trace Elements (Boron, Iron, Lead, Molybdenum, Cadmium, Lithium, Zinc and others) were measured on samples of ground water collected from 1992 to 2003. Results revealed differences and diversity in the concentration of trace elements that have been studied due to the diversity of geological crust in USA, and reported an increase in the solubility of trace elements in groundwater with pH less than 7.0, except for Molybdenum which dissolve in alkaline pH [7].

Schauss (2012) noted that water mixed homogeneously with many materials, more than any other solvent and considered an optimum medium for transfer of nutrients, and through geological mixing with components of earth's crust it equips all minerals required for our bodies' health [8].

This study was conducted to determine the chemical composition of groundwater in Al-Jadiriyah, Baghdad province to assess trace and heavy elements (Iron, Manganese, Zinc, Copper, Cobalt, Lead and Cadmium and Nickel) present in the ground water in wells Al-Jadriya.

2. Materials and Methods

Samples: Water samples were collected from eight wells, in Al-Jadriya, at area lies between Al-Mualak bridge and AL-Jadiriya bridge parallel to the path of the Tigris River (at depth from 15 to 20 meters), monthly for one year from June 2010 to May 2011, **Figure 1**.

Figure 1. Google Map shows Well sites between Al-Mualak bridge and AL-Jadria bridge across Abu Nuwas street and University Camp.

Preparation of Sample Solutions: Sterile plastic bottles (5000 cm^3), were used for collection of water. Samples were stored at 4°C through the analytical stages for keeping it from evaporation.

100 ml of water samples were taken, to which 20 ml concentrate (65%) Nitric acid was added; transferred to plastic vessel (beaker), heated in a lab-guide microwave oven at low-medium power for 10 - 15 minutes. Cooled and transferred to a volumetric flask, then rediluted to 100 ml final volume.

Trace elements (Fe, Zn, Mn, Cu) and heavy metals (Pb, Co, Ni, Cd) were measured.

Instrument: Buck Model 210/211 AAS furnished with; Air flame (for Fe and Cu) and Furnace (for Pb and Fe). The wavelengths for each element. Described in (**Table 1**).

Calibration: A linear, 2-point Calibration was made using the appropriate Matrix Blank for the Flame or Furnace and the Standards set within the CAL MAX range for each metal [9].

3. Results and Discussion

The average quarterly results of the trace elements that have been assessed in 96 water samples from eight wells in this study for the period from June 2010 to May 2011 are:

Iron: analysis revealed concentration range (0.12 to 0.29) ppm in most water samples (**Tables 2-5**). This range is less than limited values specified by WHO (0.3 ppm). Iron is an important RBC component, and its deficiency causes anemia, but its presence in the water in excess of the allowable limit (0.3 - 1.0 ppm) causes toxicity and poor taste water [10]. Manganese: trace amount range was (16.00 - 443 ppb) (**Tables 2-5**), this level is within the proper focus of this element according to the WHO guide (500 ppb) (WHO, 2011). Zinc concentrations ranged (0.13 ppm - 0.71 ppm), (**Tables 2-5**), it is less than limitations of the WHO (3.0 ppm). Zinc is essential for plant growth and human and added as soil fertilizer and as medical creams to address the shortage in patients who suffer from skin infections [10].

Results revealed absence of Copper, Cobalt, Lead, Cadmium and Nickel: in water samples, (**Tables 2-5**), so wells' water is not-polluted.

The low concentration or absence of trace and heavy elements in wells' water is related to type of soil, the Sedimentary soil in Mesopetenium region, resulting from flood plain deposit which is poor with this elements [11], and the absence of excess of fertilizers, insecticides and industrial pollution. This gives pure well's water free from pollutions [12].

Trace elements and heavy metals concentrations that included in our study did not exceed the concentrations allowed for drinking water set by the Central Organization for Standardization and Quality Control [3].

Table 1. Shows the elements and wavelengths that have been taught at the measurement (BUCK, 2006).

Element	Fe	Mn	Zn	Cu	Co	Pb	Cd	Ni
Wavelengths nanometer (ηm)	245.3	279	213.9	324.8	240.7	217	228.9	232

Table 2. Average concentration of trace elements and heavy water samples wells Jadiriya months of June July and August 2011 for eight wells.

Limited values WHO ppm	Tiger water	Well No. 8	Well No. 7	Well No. 6	Well No. 5	Well No. 4	Well No. 3	Well No. 2	Well No. 1	Unit	Element
0.3	Nil	0.29	0.25	0.17	Nil	0.18	Nil	0.12	Nil	ppm	Fe
0.5	Nil	Nil	Nil	Nil	Nil	Nil	Nil	Nil	16.00	ppb	Mn
3.0	Nil	0.32	0.25	0.13	0.22	0.24	0.15	0.25	0.23	ppm	Zn
2.0	Nil	Nil	Nil	Nil	Nil	Nil	Nil	Nil	Nil	ppb	Cu
-	Nil	Nil	Nil	Nil	Nil	Nil	Nil	Nil	Nil	ppb	Co
0.01	Nil	Nil	Nil	Nil	Nil	Nil	Nil	Nil	Nil	ppb	Pb
0.003	Nil	Nil	Nil	Nil	Nil	Nil	Nil	Nil	Nil	ppb	Cd
0.02	Nil	Nil	Nil	Nil	Nil	Nil	Nil	Nil	Nil	ppb	Ni

ppm = part per million, ppb = part per million, 1 ppm = 1000 ppb.

Table 3. Average concentration of trace elements and heavy water samples wells Jadiriya months of Sept October and November 2011 for eight wells.

Limited values WHO (ppm)	Tiger water	Well No. 8	Well No. 7	Well No. 6	Well No. 5	Well No. 4	Well No. 3	Well No. 2	Well No. 1	Unit	Element
0.3	Nil	Nil	Nil	Nil	Nil	Nil	Nil	Nil	Nil	ppb	Fe
0.5	Nil	Nil	Nil	Nil	Nil	Nil	Nil	40.00	58.00	ppb	Mn
3.0	Nil	30.00	10.00	33.00	25.00	19.00	12.00	23.00	41.00	ppb	Zn
2.0	Nil	Nil	Nil	Nil	Nil	Nil	Nil	Nil	Nil	ppb	Cu
-	Nil	Nil	Nil	Nil	Nil	Nil	Nil	Nil	Nil	ppb	Co
0.01	Nil	Nil	Nil	Nil	Nil	Nil	Nil	Nil	Nil	ppb	Pb
0.003	Nil	Nil	Nil	Nil	Nil	Nil	Nil	Nil	Nil	ppb	Cd
0.02	Nil	Nil	Nil	Nil	Nil	Nil	Nil	Nil	Nil	ppb	Ni

Table 4. Average concentration of trace elements and heavy water samples wells Jadiriya months of December January and February 2011 for eight wells.

Limited values WHO (ppm)	Tiger water	Well No. 8	Well No. 7	Well No. 6	Well No. 5	Well No. 4	Well No. 3	Well No. 2	Well No. 1	Unit	Element
0.3	Nil	Nil	Nil	Nil	Nil	Nil	Nil	Nil	Nil	ppb	Fe
0.5	Nil	Nil	Nil	33	15	70	62	33	104	ppb	Mn
3.0	Nil	100	24	20	15	10	56	40	580	ppb	Zn
2.0	Nil	Nil	Nil	Nil	Nil	Nil	Nil	Nil	Nil	ppb	Cu
-	Nil	Nil	Nil	Nil	Nil	Nil	Nil	Nil	Nil	ppb	Co
0.01	Nil	Nil	Nil	Nil	Nil	Nil	Nil	Nil	Nil	ppb	Pb
0.003	Nil	Nil	Nil	Nil	Nil	Nil	Nil	Nil	Nil	ppb	Cd
0.02	Nil	Nil	Nil	Nil	Nil	Nil	Nil	Nil	Nil	ppb	Ni

Table 5. Average concentration of trace elements and heavy water samples wells Jadiriya months of March April and May 2011 for eight wells.

Limited values WHO (ppm)	Tiger water	Well No. 8	Well No. 7	Well No. 6	Well No. 5	Well No. 4	Well No. 3	Well No. 2	Well No. 1	Unit	Element
0.3	Nil	Nil	Nil	Nil	Nil	Nil	Nil	Nil	Nil	ppb	Fe
0.5	Nil	Nil	Nil	361	153	92	443	84	60	ppb	Mn
3.0	Nil	150	30	20	30	50	110	50	710	ppb	Zn
2.0	Nil	Nil	Nil	Nil	Nil	Nil	Nil	Nil	Nil	ppb	Cu
-	Nil	Nil	Nil	Nil	Nil	Nil	Nil	Nil	Nil	ppb	Co
0.01	Nil	Nil	Nil	Nil	Nil	Nil	Nil	Nil	Nil	ppb	Pb
0.003	Nil	Nil	Nil	Nil	Nil	Nil	Nil	Nil	Nil	ppb	Cd
0.02	Nil	Nil	Nil	Nil	Nil	Nil	Nil	Nil	Nil	ppb	Ni

From above results, we can conclude that wells water can be used under special management if no other problems are presented.

References

[1]   Buringh, P. (1960) Soils and Soil Conditions in Iraq. Ministry of Agriculture, Baghdad.

[2]   World Health Organization (WHO) (2011) Guide Lines for Drinking-Water Quality. 4th Edition, Geneva, 30-120.

[3]   Central Organization for Standardizations and Quality Control (2001) Standard Criteria for Drinking Water. IQS/417/2001. Ministry of Planning, Republic of Iraq.

[4]   Boyd, C.E. (2000) Water Quality an Introduction. Kluwer Academic Publishers, Kluwer, 219.

[5]   Pendias, A.K. and Pendias, H. (2001) Trace Elements in Soil and Plants. 3rd Edition, CRC Press, Boca Raton, 25-41.

[6]   Al-Dulaimi, A.M.C. (2011) Effect of Sewage Water of Fallujah City on the Chemical Pollution of Euphrates Water, Soil and Plant. Ms.C. Thesis, Collage of Agriculture, University of Al-Anbar, Ramadi.

[7]   Ayotte, J.D., Gronberg, J.A.M. and Apodaca, L.E. (2011) Trace-Elements and Radon in Groundwater across the United State, 1992-2003. National water-Quality Assessment (NAWQA) Program, US Geological Survey Scientific Investigations Report .

[8]   Schauss, A.G. (2012) Minerals and Human Health the Relation for Optimal and Balanced Trace Element Levels. http://www.traceminerals.com/research/humanhealth

[9]   Buck Scientific 205 Atomic Absorption Spectrophotometer (2006) Operator's Manual. 67.

[10]   Jinwal, A., Dixit, S. and Malik, S. (2009) Some Trace Element Investigation in Ground Water of Bhopal and Sehore District in Madhya Pradesh: India. *J. Apple. SCI. Environ. Manage*, **13**, 47-50.

[11]   Al Ammar, H.A.S., Jalal, H.A.A. and Jasem, A.J. (2008) Study of Heavy Metals Concentrations in Ground Water in Hilla Government. College of Science, University of Babel.

[12]   Radhy, S.H., Mosa, A., Qader, Z. and Muhamed, A. (2011) Study of Heavy and Trace Elements in Iraqi Soils. Ministry of Environment, Department of Natural Environmental Systems.

5

Development of a Strategic Planning Model for a Municipal Water Supply Scheme Using System Dynamics

Adelere E. Adeniran[1*], Olufemi A. Bamiro[2]

[1]Civil & Environmental Engineering Department, University of Lagos, Lagos, Nigeria
[2]Mechanical Engineering Department, University of Ibadan, Ibadan, Nigeria
Email: [*]engrea@yahoo.com, femi.bamiro@skannet.com

Abstract

This paper reports the development of a system dynamics model for the strategic planning of a municipal water supply scheme. The model is capable of handling the critical variables that impact on the operations of a water supply scheme. The model was developed using the dynamo programming software to handle the process dynamics. Other supporting softwares—Fortran, Clipper 5, and Epiglue are used to handle data input, analysis and presentation of results in a user-friendly environment. The model was validated by applying it to the simulation of the University of Ibadan Water Supply Scheme for which extensive data of operations were collected over a period of ten years. The results obtained were found to be realistic and useful for the planning of plant operations. It is concluded that the model developed is sensitive to variable changes and has the capability of answering various operational "what-if" questions. As a strategic planning tool, the present model generates, for different scenarios, important operational information for the management of the water supply scheme. It is concluded that SD model is a useful tool to assist water managers and policy makers in making decisions and evolving strategic planning for water supply.

Keywords

Municipal Water Supply, System Dynamics, Strategic Planning, Policy Decisions

1. Introduction

The city of Ibadan has been faced with water scarcity since the late 1980s. The scarcity of water in the city af-

---

[*]Corresponding author.

fected the University of Ibadan (UI) culminating in a fire incident that gutted a female undergraduate hostel. The management and operation of a Water Supply Scheme in an African metropolis is a complex and challenging problem that has cultural dimensions and demands a high level of strategic planning. A tool that can facilitate the performance of these functions is a step in the right direction [1].

The management of the scarce water requires the development and operation of a model that can handle the complex dynamics that is involved [2]. In this paper, a System Dynamic (SD) model is used to capture the dynamics and the structure of the UI water supply system. The developed model was tested as a tool for making strategic planning decisions using simulations to replicate various scenarios. The objective of such a Strategic Planning Model for a Municipal Water Supply Scheme (SPMWSS) was to develop a strategic planning tool that would give the water managers, planners, policy and decision makers, flexibility in assembling and testing water management strategies for the university and similar settlements. A strategy may include structural measures to increase supply capacity as well as managerial options to manage water supply, reduce costs and to allocate water supply to various demand units at varying quantities and time [3].

The simulation system is designed with the appreciation that there are several limitations on the processing and computing abilities of human managers and decision makers. A model for water supply management must necessarily focus attention on the flow of information in a complex water supply management system, and the use of such information in the decision making process. It should also focus attention on the role that simulation can play in gaining insights into current and future water management policies and strategies [4].

Three types of information can be used in the development of a strategic planning model: *mental, written and numeric* [5] [6]. In the development of the SPMWSS model, the *mental information* relied on the mental data base of the authors and those of their colleagues in the water unit of the University and it reliably captured the structure and policies governing management of the water supply scheme in UI. On the other hand, *written data* on water supply problem and modelling strategies abound. Several articles and conferences have tried to focus attention on paradigm shift and the need to apply modelling techniques to strategic planning in the water supply industry. The principles of SD are well suited for modeling and application to water resources and environmental problems [7]-[9]. SD is as an emerging tool with great potentials for improved understanding of environmental systems [10]. *Numerical data* used in building the model were collected and reviewed. For our problems, these included data from 1991 to 2003 kept by the UI water unit and unpublished reviews from consultants engaged at different times for improvement to the university's water supply.

2. Concept, Structure and Development of the SPMWSS

The simulation model was formulated base on System Dynamics theory. System Dynamics, as an aspect of system theory, is a method for understanding the dynamic behavior of complex systems. The governing equations represented by finite difference expressions used for modeling different elements in a system are solved using standard numerical schemes [11]. System Dynamics modelling has attracted considerable attention over the past few decades. In a particular sense, System Dynamics is concerned with the use of models and modelling techniques to analyse complex systems and policy issues with a view of getting the right combination of scenarios to accomplish an efficient strategic planning for the system being modelled [3].

In the SPMWSS model, DYNAMO, FORTRAN, CLIPPER 5 and EPIGLUE programming languages are combined to develop a user-friendly and menu-driven environment to drive the model. The simulation model consists of four major sectors.

The finance module includes details of the capital, operational and maintenance financial analysis. It is further sub-divided into (i) total cost in a 12-month cycle which can be used for budgeting and (ii) production and unit cost analysis which can be used for tariff setting mechanisms. The production module includes the model representation of population, per capita demand and supply and demand management options. The operation and maintenance module captures the pumping operations, availability, spare parts and maintenance schedules. It also includes store and inventory schedules. The distribution module lists the various water uses and demand nodes and their usage potentials. It incorporates the reservoirs and the pumping stations. Each module is a standalone but is interlinked with other module. Each of the modules also represents an aspect of water management from the point of view of sectoral interdependency (**Figure 1**).

The four modules are then housed in the SPMWSS model as interactive and interdependent components for the strategic planning of a Municipal Water Supply constituting the global system.

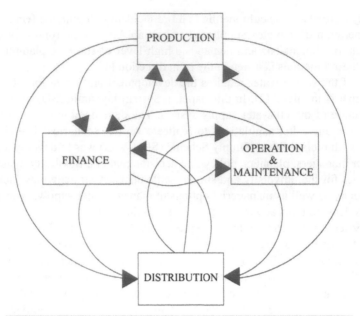

Figure 1. Interconnectivity of the major components of the SPMWSS model. Source: Adeniran (2007).

The concept of the SPMWSS model is as illustrated in **Figure 2**. The total model is segregated into three stages *viz.*: (*i*) *Supporting Data Interface*, (*ii*) *Main SPMWSS Simulation Model and* (*iii*) *Report Generator Systems*. The three-stage model concept is based on the identified interdependencies. In the SPMWSS, the DYNAMO language was adopted to model the various components and variables identified. The Equations are based on the mathematical philosophy of System Dynamics, the interconnectivity of the Water Supply System and the Syntax of DYNAMO modelling. The complete DYNAMO program listing (UIWATER.DYN) for the SPMWSS and the exogenous data file (UIWATER.ASC) is presented in [3].

3. Model Validation

The SPMWSS model is validated based on the rendition that the model must have the ***ability to replicate the real life system*** being modeled under the same parametric conditions. It is the validation of the basic underlying processes of the model (***Design Validity***). Selected endogenous variables, which are representative of the water supply system's behaviour, which have been collected from 1991 to 2003, are used to validate the model over a 12-month planning horizon each. Comparison of the model output with actual historical data over the same time period provides the required validation.

Figure 3 shows the comparison of the Actual Production for the years 1992 to 2002 under the real life operating conditions as available in the Water Works records as compared with the simulated outputs of the SPMWSS model for the same period. A comparison of simulated versus actual data shows that the variance is less than 5% and hence considered within reasonable limits.

Figure 4 compares the Actual Monthly production as available in the Water works records with the SPMWSS simulated Production under the same operating conditions for the year 2002. It is concluded from the two examples that the model replicates the system fairly well.

4. Application of SPMWS: Scenario Investigations

Since the primary objective in building SPMWSS is the conduct of scenario experiments for strategic planning, considerable thought is given to the types of experimental features that must be built into SPMWSS to ensure effective and efficient scenario simulations can be carried out. The utilization of SPMWSS is oriented toward scenario experiments to production, finance and maintenance situations that will assist Management in short term and long-range and strategic planning of the water supply system. Scenario investigations allow the model users to ask "what-if" questions. The following strategic "what-if" questions were investigated using the model

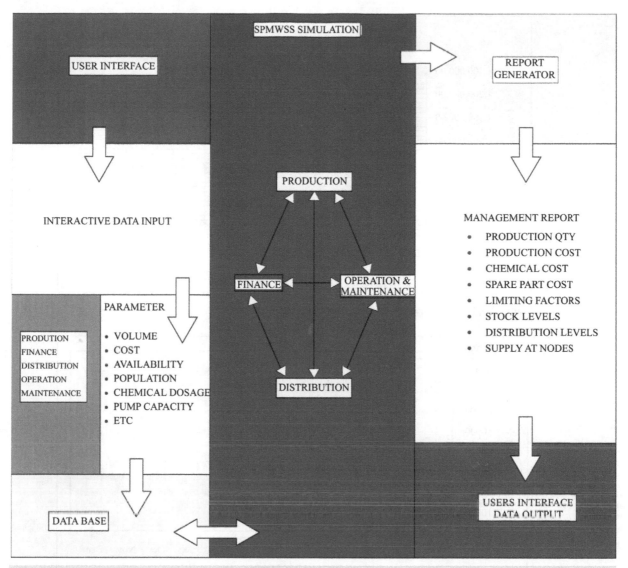

Figure 2. Concept of the SPMWSS model. Source: Adeniran (2007).

and the results presented.

Scenario 1: What if population should increase but per capita demand is constant?

Scenario 2: What if the plant efficiency should depreciate?

Scenario 3: What if there is a decline in raw water supply?

Scenario 4: What if the cost of energy (fuel and power) should increase?

Scenario 5: What if the plant efficiency should decrease?

Scenario 6: What if cost of raw water input from Eleyele intake (one of the raw water intake point), should increase?

Scenario 7: What if level of maintenance of the production plant should increase in order for production to remain constant?

Scenario 8: What if certain chemical dosage is applied, how much chemical?

Scenario 9: What if the production budget is required?

Scenario 1: Population Increase but Constant per Capita Demand

The simulation results of the above scenario are presented in **Figure 5**. The population is varied from 20,000 to 55,000 while the per capita water demand is kept at the WHO minimum of 110 litres. The cumulative annual productions are compared with the cumulative required production for each of the simulated situation.

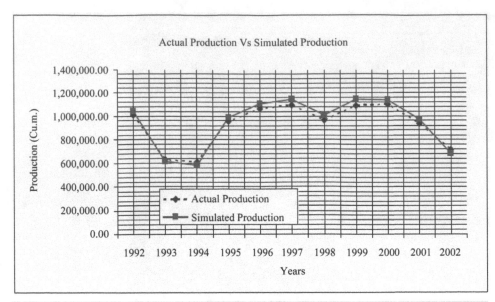

Figure 3. Chart of actual production vs. simulated production.

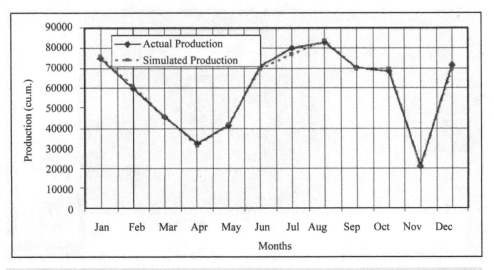

Figure 4. Year 2002 actual production vs. simulated production.

It is observed from **Figure 5** that the Water Supply Scheme is capable of meeting the per capital demand of 110 litres per person per day up to a population of about 35,000 after which it can no longer cope.

Scenario 2: Effect of Depreciating Plant Efficiency

One of the major factors militating against the ability of a water supply scheme to meet with the demand of the consumers is equipment depreciation and/or lack of maintenance resulting in decrease in Plant Efficiency. The effect of plant efficiency on water production is examined vis-à-vis the demand for water by the population. For this case, the population and per capita demand are kept constant while the plant efficiency is varied. The observations are presented in graphical form as **Figure 6**.

It is observed from **Figure 6** that the demanded quantity of water is the same as the quantity of water produced when the efficiency of the plant is higher than 65%. In other words, the plant is capable of meeting the water demand of the population provided the efficiency is not below 70%. However at lower efficiency, the plant would not be able to cope with the demanded water by the population. It should be noted that for this scenario investigation, the population and the per capita demand are kept constant.

The strategic planning implication of this is that regular maintenance activities should be carried out to ensure that the efficiency of any the components of the system does not drop below 70%. This can be achieved by

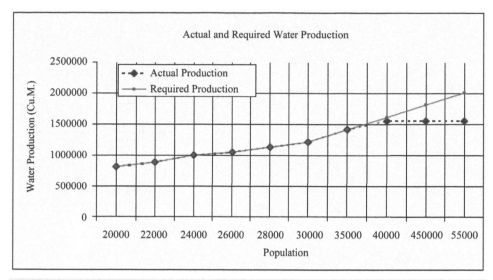

Figure 5. Actual production compared with water demand with varying population.

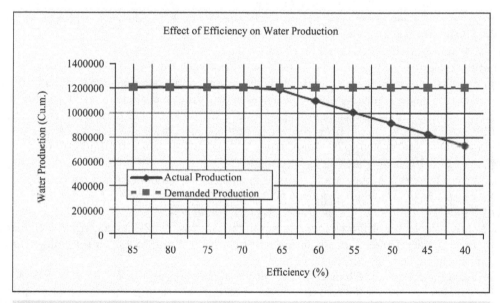

Figure 6. Effect of plant efficiency on water production.

conducting regular reliability tests on the equipment.

Scenario 3: Declining Raw Water Supply

At present, the UI Water supply system modeled obtains raw water from both Eleyele and Oba Dam. While Eleyele is an external supply source, the Oba Dam is an internal supply source. A situation that can arise is for Eleyele source may not be available or the Oyo State Government who is the owner of the dam could come up with an exclusion policy and the University is stopped from the abstraction of raw water from Eleyele Dam. Presently 60% of the University's raw water is from Eleyele. In order to simulate the above condition, the scenario in which the raw water input of Eleyele dam is reduced gradually from 60% to 0% is simulated. This is done by reducing the hours of operation of the Eleyele pumps form 20 hours a day to zero hours a day. For this scenario, the population is assumed to be 30000 and per capita demand is 110 litres. The effect of this scenario on water production is as shown in **Figure 7**.

From **Figure 7**, the limiting factor becomes 1.0 when the raw water contribution from Eleyele becomes 12.5%. Any further reduction implies that the water demand of the population cannot be met. In fact, at 0% contribution from Eleyele, the limiting factor is 0.77. In other words, Oba Dam raw water source alone can cope

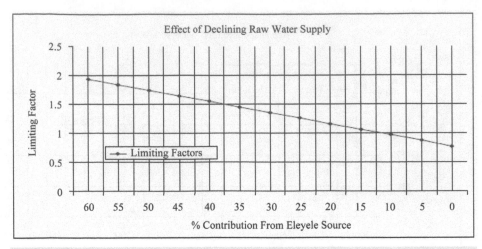

Figure 7. The effect of declining raw water supply on production.

with only 77% of the water demand of the population.

The strategic planning implication is for Management to plan ahead and look for alternative source of raw water. For instance, the long proposed Ona river dam (see **Figure 8**) could be a feasible option to counteract the effect of declining raw water supply either from Eleyele or Oba dam.

Scenario 4: Increase in Cost of Fuel and Energy

Some of the major costs of production include the cost of Fuel and Energy (PHCN). From 1992 to 2005, the Nigerian Economy witnessed many fluctuations in the cost of energy and fuels. The effects of these changes are examined using the model. First, the effect of changes in the cost of fuel alone is examined, then the effect of the cost of energy (PHCN) alone is examined, and then the combined effects of these changes are examined together. The simulation results of these effects on the cost of Production and the Unit cost of production are presented in the following sections.

(a) Effect of Fuel Price on Unit Cost of Production

Table 1 shows the effect of the various changes in the petroleum product prices between 1992 and 2005. It is seen that the unit cost of production rose from N55.68 to N97.17.

(b) The Effect of Cost of Energy on Unit Cost of Production

Table 2 shows the effect of the various changes in the unit cost of (PHCN) energy N/kWh between 1992 an 2005. It is seen that the unit cost of production rose from N59.12 to N85.48. This is a single factor simulation while the other factors are kept constant.

(c) The Combined Effect of Changes in Cost of Fuel and Energy

Usually, the changes in a factor lead to changes in other factors in real life. Especially, whenever there is a change in the cost of petroleum products there is usually a correspondence change in the cost of energy and vice versa. **Figure 9** shows the changes.

It is seen that the cost of production increases with increases in cost of fuel and energy. This result is expected, but with the SPMWSS model, it is now possible to quantify the effect of any slight changes in any of these important factors. It should be noted that these results are for a fixed population and fixed production quantity. The model is also capable of examining the effect of the economy of scale with varying production quantities [1], [2].

Scenario 5: Declining Plant Efficiency on Unit Cost of Production

One of the major factors in cost consideration is the efficiency of the production plant. The reduction in the plant efficiency does not only affect the production output it also affect the cost of production. In order to observe the effects of the efficiency on the cost of production, the efficiency is gradually reduced from the normal operating efficiency of 85% to 5%. The results of simulation are presented in the **Figure 10**.

It is seen from simulation results that the unit cost of production increases exponentially with linear declining of Plant Efficiency. It is obvious that it is not wise to operate the plant at low efficiency. The strategic plan implication of this is for adequate attention to be placed on maintenance such that efficiency is not allowed to drop. Any equipment that is operation below 70% efficiency should either be overhauled or replaced.

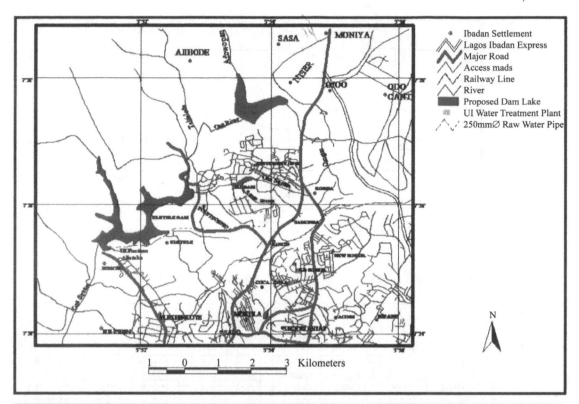

Figure 8. Map showing the locations of water supply reservoirs to university of Ibadan. Source: Works and Maintenance Department, University of Ibadan.

Table 1. The effect of changes in fuel prices on cost of production.

PMS Price =N=	0.70	3.23	11.00	25.00	20.00	22.00	26.00	34.00	49.90	50.50
AGO Price =N=	0.50	2.50	7.50	15.00	12.50	22.00	27.50	42.00	52.00	65.00
Unit Cost =N=	55.68	57.09	60.80	66.65	64.66	70.10	73.63	82.53	89.99	97.17

Table 2. Effect of changes in unit cost of energy on unit cost of production.

Year	Changing PHCN Price =N=/kWh	Corresponding Unit Cost =N=/cu.m
1992	1.50	59.12
1993	3.00	62.89
1998	4.50	66.65
1999	6.00	70.43
2000	7.50	74.18
2001	7.50	74.18
2005	12.00	85.48

Scenario 6: Cost of Raw Water Input from Eleyele Intake

Raw water is a major production input. At present, the University abstracts raw water from two sources, Eleyele Dam belonging to the Water Corporation of Oyo State (WCOS) and the Oba dam which belongs to the University. Payment of raw water to the WCOS has been very controversial from the inception of the University Scheme in 1990. The WCOS has insisted on payment for raw water and has increased the raw water tariff from N1.50, to N3.0 and now N6.0 per cu.m. In order to examine the effect of increase of raw water input from

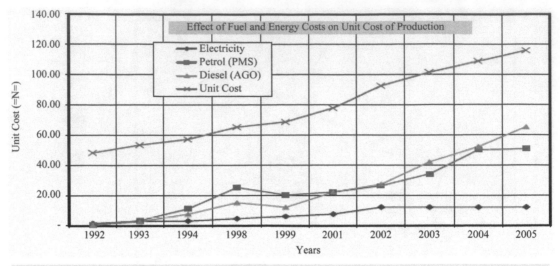

Figure 9. Effect of fuel and energy costs on unit cost of production.

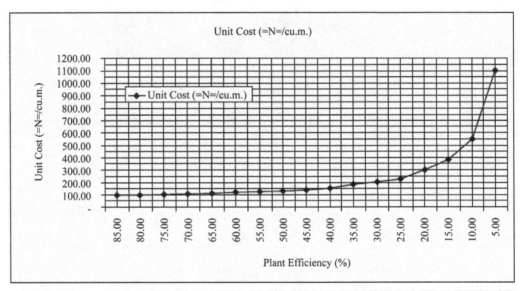

Figure 10. The effect of declining plant efficiency on unit cost of production.

Eleyele, a scenario is created in which the cost of raw water from Eleyele is increased from N3.0 to N100.00. The results are presented in **Figure 11**.

It is seen that the production cost increases as the cost of raw water input from Eleyele increases. The Strategic Planning implication is evident from this graphic example: Suppose the WCOS decides to increase the cost of raw water from the present N6.00 per cubic meter to say =N= 100.00 per cubic meter, the unit cost of production will change from N92.11/m^3 to N234.53/m^3. This type of information can be used as argument against an arbitrary increase in the cost of raw water. It can also be used to justify the need for the University to look for its own internal sources of raw water.

Scenario 7: Increase in Level of Required Maintenance to Sustain Expected Production

The plant capacity is usually fixed relative to the level of maintenance input. If the plant is to be expected to have a constant production level, then the maintenance activities on the plant must be kept regular. Managers and operators are often requested to keep the plant running without corresponding supply of spare parts to ensure the optimum performance of the plant. Effective and cost effective maintenance also involve keeping reasonable level of staff for the running of the plant as well as evolving control policies as to the level of fuel consumptions and running cost expenses. Strategic planning in the area of maintenance, therefore, places primary emphasis on cost control with respect to the level of maintenance for constant production output.

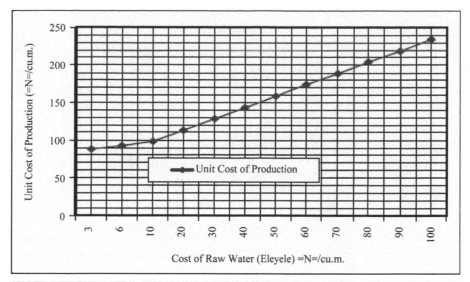

Figure 11. Effect of cost of raw water from Eleyele on unit cost of production.

The required level of maintenance of a system is a function of the maintenance input into the system. A new system requires a relatively low level of maintenance whereas a system that is tending towards unserviceability requires a very high level of maintenance. The aggregate of spare part usage, manpower attendance and down times are captured in terms of percentage. The required level of maintenance is varied from 10% to 100% while the production parameters are kept constant. The unit cost of production is observed for each of the simulated situation. The results are presented in **Figure 12** and **Table 3**.

It is seen that the unit of production cost increases as the required level of maintenance increases. It is noted that when the level of maintenance is 0%, the unit cost is =N= 97.46 whereas when the level of maintenance required rises up to 100% i.e. requiring shut down maintenance, the unit cost of production is =N= 140.96.

It is also noted from **Table 3** that if the level of required maintenance should increase from 0% to only 10%, the University has to look for an additional sum of =N= 5,000,000.00 in order to maintain production. The situation becomes worse with higher level of required maintenance. If the level of required maintenance is allowed to reach 100%, the university must look for as much as =N= 52,535,000.00 for turnaround maintenance. The strategic planning implication is that management must evolve and sustain a maintenance policy that ensures routine maintenance in order to limit the production cost. Under no condition must equipment be allowed to depreciate to the level of breakdown maintenance.

Scenario 8: Total Chemical Usage Given Chemical Dosage

As an Expert System, the SPMWSS can be used as an oracle to tell the total quantities of chemical expected to be used if the level of chemical dosages at each period of the planning horizon is known. The model can thus be used to monitor chemical usage of the treatment plant and thus prevent and or detect theft. In this example, Alum alone is used. However, other chemicals can be equally monitored. The dosage of Alum is varied from 80 ppm to 180 ppm based on the turbidity profile of the raw water which varies from season to season. The best Alum dosage is about 80 ppm while the worse is about 180 ppm at the peak of the raining season. **Table 4** shows the expected Total Annual Alum usage at a given dosage. This information can be used to intelligently figure out the quantity of Alum that Management should purchase if the average dosage in a year is known.

Scenario 9: Expert System for Production Budget

One of the attributes of the SPMWSS model is its budgeting capability. In a public establishment, budgets are often required within a short time frame. Such budgets are usually prepared using the rule of the thumb. SPMWSS model is an Expert System for preparing production budget because all the financial components for preparing a production budget have been built into the model. In actual sense, the total cost under certain conditions can be given as a model output as shown in **Table 5**. It is seen from this table that cumulative costs of each of the vital production item for the simulated condition are presented. Thus it is known that about =N= 6,161,200 is required for Chemicals while about =N= 511,000.00 is required for spare parts and a total of =N= 117,716,000 is the total money required for the water sector during the planning year.

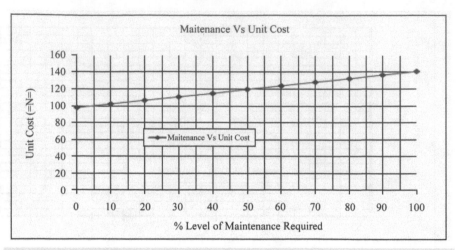

Figure 12. Effect of increased level of maintenance on unit cost of production.

Table 3. Effect of increased level of maintenance on cost of production.

% Level of Required Maintenance	Total Production Cost =N=:k	Unit Cost of Production =N=:k
0	117,716,000.00	97.46
10	122,977,000.00	101.82
20	128,218,000.00	106.16
30	133,468,000.00	110.51
40	138,711,000.00	114.85
50	143,976,000.00	119.21
60	149,242,000.00	123.57
70	154,486,000.00	127.91
80	159,727,000.00	132.25
90	164,979,000.00	136.59
100	170,251,000.00	140.96

Table 4. Total alum usage given dosage level.

Alum Dosage (ppm)	80	90	100	110	120	130	140	150	160	170	180
Alum Usage (Kg)	96,627	108,702	120,780	132,840	144,960	157,020	169,080	181,121	193,260	205,320	217,380

5. Conclusion

In this paper, the development of a computer based strategic planning model for a water supply scheme has been reported. The model is developed using the DYNAMO modeling software. However, in order to create a user-friendly environment, other application programs such as FORTRAN, CLIPPER 5 DBASE, EXCEL and EPIGLUE have also been used in an integrated manner. The model was validated using existing data that spanned ten (10) years from the University of Ibadan Water Supply Scheme. The model was then used to carry out scenario experimentations for strategic planning. The results presented indicate that an SPMWSS model has been successfully developed to handle the operation of the University of Ibadan Water Supply Scheme. The potential of greater understanding of a system through the system dynamics approach, which is still in its infancy, is shown to be not only feasible but also exciting. The model is found useful to managers and planners for its strategic planning capability. The SPMWSS model integrates the production and financial elements; in addition to

Table 5. Cumulative costs of various production item.

PERIOD	TOTAL EXP.	CHEN. COST	SPARE PARTS	INCIDENTALS	SALARY	POWER COST	TRANSPORT
₦	₦	₦	₦	₦	₦	₦	₦
1	9,723,000.00	454,400.00	5000.00	5000.00	940,000.00	3,081,000.00	4,067,000.00
2	18,901,000.00	879,500.00	16,000.00	10,000.00	1,880,000.00	5,963,000.00	7,872,000.00
3	28,645,000.00	1,333,900.00	42,000.00	15,000.00	2,820,000.00	9,044,000.00	11,939,000.00
4	38,120,000.00	1,773,800.00	74,000.00	20,000.00	3,760,000.00	12,026,000.00	1,587,500.00
5	47,879,000.00	2,264,000.00	79,000.00	25,000.00	4,700,000.00	15,107,000.00	19,942,000.00
6	57,396,000.00	2,773,200.00	84,000.00	30,000.00	5,640,000.00	18,089,000.00	23,878,000.00
7	68,156,000.00	3,370,800.00	260,000.00	35,000.00	6,730,000.00	21,125,000.00	28,559,000.00
8	7,891,600.00	3,968,400.00	436,000.00	40,000.00	7,820,000.00	24,161,000.00	33,240,000.00
9	88,497,000.00	4,546,900.00	436,000.00	45,000.00	8,760,000.00	27,143,000.00	37,176,000.00
10	98,370,000.00	5,144,500.00	448,000.00	50,000.00	9,700,000.00	30,224,000.00	41,243,000.00
11	107,930,000.00	5,671,000.00	479,000.00	55,000.00	10,640,000.00	33,206,000.00	45,179,000.00
12	117,116,000.00	6,161,200.00	511,000.00	60,000.00	11,580,000.00	36,287,000.00	49,246,000.00

achieving true total cost perspective, this particular capability introduces the potential flow paths from production to consumption. The SPMWSS provides a mechanism wherein the modeled system can be examined under scenario experiments. In addition, many other aspects of system management as they relate to water supply can be tested in an effort to evaluate existing policies and principles.

References

[1]   Adeniran, A.E. and Bamiro, O.A. (2010) A System Dynamics Strategic Planning Model for a Municipal Water Supply Scheme. *Proceedings of the International Conference of the System Dynamics Society*, Seoul, 25-29 July 2010.

[2]   Adeniran, A.E. (2014) Application of System Dynamics Model in the Determination of the Unit Cost of Production of Drinking Water. *International Journal of Water Resources and Environmental Engineering*, **6**, 183-192. http://dx.doi.org/10.5897/IJWREE2013.0479

[3]   Adeniran, A.E. (2007) Development of a System Dynamics Strategic Planning Model for a Municipal Water Supply System. Ph.D. Thesis, University of Ibadan, Ibadan.

[4]   World Bank (1993) A Strategy for Managing Water in the Middle East and North Africa, IBRD. World Bank, New York.

[5]   Forester, J.W. (1961) Industrial Dynamics. MIT Press, Cambridge.

[6]   Forrester, J.W. (1992) Policies, Decisions and Information Sources for Modeling. *European Journal of Operation Research*, **59**, 42-63. http://dx.doi.org/10.1016/0377-2217(92)90006-U

[7]   Fletcher, E.J. (1998) The Use of System Dynamics as a Decision Support Tool for the Management of Surface Water Resources. *First International Conference on New Information Technologies for Decision Making in Civil Engineering*, Montreal, 11-13 October 1998, 909-920.

[8]   Ford, A. (1999) Modeling the Environment—An Introduction to System Dynamics Modeling of Environmental Systems. Island Press, Washington DC.

[9]   Deaton, M.L. and Winebrake, J.I. (1999) Dynamic Modeling of Environmental Systems. Springer-Verlag, New York.

[10] Huang, G.H. and Chang, N.B. (2003) Perspectives of Environmental Informatics and Systems Analysis. *Journal of Environmental Informatics*, **1**, 1-6. http://dx.doi.org/10.3808/jei.200300001

[11] Bamiro, A.O. (1995) Electronic Oracle—Seeing through Computer Models. *Post Graduate School Interdisciplinary Discourse*, University of Ibadan, Ibadan.

6

Water Resources Conflict Management of Nyabarongo River and Kagera River Watershed in Africa

Telesphore Habiyakare[1,2], Nianqing Zhou[1*]

[1]Department of Hydraulic Engineering, Tongji University, Shanghai, China
[2]College of Science and Technology, University of Rwanda, Kigali, Rwanda
Email: t.habiyakare@yahoo.com, *nq.zhou@tongji.edu.cn

Abstract

In the process of exploiting and using water resources of river basin, the benefit conflict problems among upper and lower river districts and among different departments restrict to sustainable exploiting and using water resources of river basin. In this paper, the water resources conflict management of Nyabarongo River and Kagera River watershed is studied. The Nyabarongo is a major river in Rwanda, begins in Nyungwe Forest, and flows up to the north-western part of the country, then down through the center to the south-east, eventually forming the main tributary of the Kagera River watershed, the main affluent of Lake Victoria, which drains into the Nile River. The basin is shared among 11 riparian states. This trans-boundary character of the Nile presents a great challenge of water conflicts; national interests have historically been promoted at the expense of regional interests. The framework of this paper is as follows: the water resources bulletin is firstly described, and then the cooperation and regional conflicts are discussed; finally a sustainable framework for governing the water resources is proposed to meet water management in riparian states.

Keywords

Water Management, Water Conflicts, Nile River, Kagera Watershed

1. Introduction

Rwanda is a mountainous area with an altitude ranging between 900 m and 4507 m, and has a tropical temperate

---

*Corresponding author.

climate due to its high altitude. The average annual temperature ranges between 16°C and 20°C, without significant variation. Rainfall is abundant although it presents some irregularities. The average arable surface area available is about 0.60 ha per household use. This causes overexploitation of available land, which is often accompanied by agricultural malpractices with disastrous consequences on land resources and on environment in general [1]. The degradation of the natural environment is particularly linked with soil erosion that affects the important portion of agricultural land. Generic impacts of erosion are numerous: increasing sedimentation on land cultivated downhill from eroded plots; loss of soil fertility by leaching arable lands; risk of destruction of crops and sand banks which are particularly high in marshlands and valleys; risks of landslides [2].

Rwanda possesses a relatively big quantity of surface water and underground water: rivers, lakes and marshlands occupy a surface area of 211,000 ha, which is about 8% of the national territory (lakes: 128,000 ha, rivers: 7260 ha and marshlands: 77,000 ha). The outflow of the renewable underground resource is estimated at 66 m^3/s, out of which 9 m^3/s is produced by 22,000 known sources. In general, too little information is available especially on underground water aquifers [3]. Rwanda counted 11,689,696 inhabitants with a surface area of 26,338 km^2 during August 2012 national census with an annual growth rate of 3%, which means: the population density of 419.8 inhabitants per km^2 and which is one of the highest densities on African continent. The gross domestic product (GDP) of Rwanda is dominated by the agricultural sector [4]. This has a great impact on water use in general, especially in Kagera watershed.

2. Methods

The aim of writing this paper is to review information about Kagera watershed, as well as proper management of the users. The paper will first concentrate on introduction and brief background of Kagera Basin, then go on the situation of Kagera watershed and its boundaries. This paper will also focus on Regional River management, water resources bulletins across Nyabarongo and Kagera River including water use and demand projection, Co-operation and Regional water conflicts in Nile Basin. Finally this study will propose the mechanism for sustainable water resources management of the Kagera watershed and the commitment of the Nile riparian countries to foster co-operation and pursue jointly the sustainable development and management of Nile water resources for the benefit of all.

3. The Study Area Description and Regional River Management

3.1. Geographical Location

Located in the Great Lakes Region of Africa, the Kagera River Basin covers an area between the immense Lakes of Victoria, Tanganyika and Kivu. The Kagera River Basin is shared by Rwanda, Burundi, Uganda and Tanzania and flow to The Lake Victoria and drains to Nile River as shown in **Figure 1**.

The Kagera River drains a basin area of 59,800 km^2 distributed among the countries of Burundi (22%), Rwanda (33%), Tanzania (35%) and Uganda (10%) [5] as shown in **Table 1**.

3.2. Kagera Sub-Catchments

The Kagera River has its main sources in north-eastern side of Congo Nile Divide in Burundi (Ruvubu) and in the western highlands of Rwanda (Nyabarongo). It stretches about 800 km from its remotest source in the Virunga region in Rwanda to its outlet on the western shores of Lake Victoria in Uganda. The main tributaries of the Kagera River are the Ruvubu River gathering the waters from Burundi and the Nyabarongo River flowing from Rwanda.

3.2.1. The Ruvubu River

The Ruvubu River rises in the southern part of the Congo-Nile Divide in the tropical rain forest of Burundi in the province of Kayanza. Its head lies in the Kibira National Park at about 2000 metres a.s.l. and traverses about 350 km to its confluence with the Kagera River on the border between Rwanda and Tanzania [6].

3.2.2. The Nyabarongo River

The Nyabarongo River (**Figure 2**) flows over 300 km from its source in western Rwanda southwards to its outlet to Lake Rweru in south-eastern Rwanda along the border with Burundi. Its main tributary is Kanyaru River

Figure 1. Kagera River Basin location.

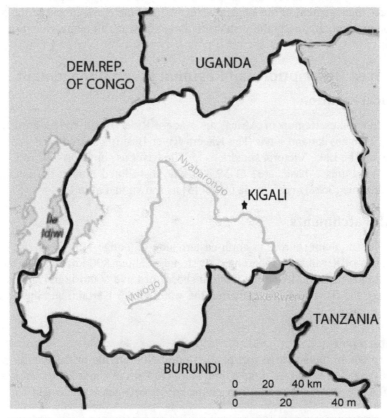

Figure 2. Map of Rwanda showing location of Nyabarongo River, which flows first north, then southeast past Kigali, then east along southeastern border of Rwanda.

Table 1. Distribution of Kagera Basin in riparian countries.

Country	Surface Area (Km²)	Total Area (Km²)	% of Kagera River Basin
Uganda	241,000	5800	10
Rwanda	26,340	19,900	33
Tanzania	945,100	20,800	35
Burundi	27,834	13,300	22
Total	**1,240,274**	**59,800**	**100**

Source: Kagera basin organisation development program, final report, 1982.

that flows from the highlands of Nyungwe National Park on the Congo-Nile Divide in Ruhengeri province along the border between Rwanda and Burundi until the junction with Nyabarongo at about 50 km south of Kigali after its turn to the mainland in Rwanda [7]. From that confluence, the Nyabarongo River flows eastwards through swampy valleys and small lakes in the lowlands of Bugesera-Gisaka in south-eastern Rwanda. From the Lake Rweru outlet, the Nyabarongo River changes the name to kagera and meanders through a swampy terrain for about 60 km and meets the Ruvubu River flowing through the Tanzanian uplands. Downstream the Kagitumba junction (which marks the border between Uganda and Tanzania), the Kagera changes direction and trends eastwards for 260 km to Lake Victoria.

3.3. River Basin Management

3.3.1. Kagera Basin Organisation (KBO)

Since 1971, numerous studies have been carried out in order to get a comprehensive development of the Kagera River Basin. In 1976, an Indicative Basin Plan provided a good Compendium of potentialities of the Kagera Basin. After many experts and officials missions throughout the basin, the four riparian countries took a political commitment to jointly exploit the water-based natural resources in order to develop their economies. Numerous conferences of donors and multidisciplinary studies were carried out and a Master Plan drawn up. The sectors of agriculture, energy, transport, communications and human resources development were chosen to be the priority sectors [8].

3.3.2. The Nile Basin Initiative (NBI)

The Nile River is one of the world's greatest assets running through 11 countries: Burundi, DRC, Egypt, Eritrea, Ethiopia, Kenya, Rwanda, South Sudan, The Sudan, Tanzania and Uganda. The region encompasses an area of 3 millions·km² and the countries of the Nile serve as a home for to an estimated 300 millions of people. The Nile River traverses 6700 km from its remotest source in the highlands of Burundi to the delta at the Mediterranean Sea in Egypt. As most other sub-Saharan African countries, the Nile riparian countries are facing poverty, political instability, rapid population growth, environmental degradation, to name but a few. From the economic perspective, the World Bank had characterized the Nile basin as a zone where "nothing flows". For the feasibility of the joint efforts to work together for the people of the Nile, the riparian governments established a cooperative framework called the "Nile Basin Initiative" with a demanding target of "achieving sustainable socio-economic development through the equitable use of, and benefit from, the common Nile Basin water resources" [9].

4. Water Resources Bulletin across Nyabarongo and Kagera River

Rwanda has a dense hydrographic network. Lakes occupy an area of 128,190 ha; rivers cover an area of 7260 ha and water in wetlands and valleys a total of 77,000 ha. The country is divided into two major water basins, divided by a water divide line or ridge called Congo Nile Ridge. To the west of this line lies the Congo River Basin which covers 33% of the national territory and which receives 10% of the total national waters. To the East lies the Nile River Basin whose area covers 67% of the territory delivers 90% of the national waters. Waters of the Nile River Basin flow out the country through the Kagera River, the main tributary to Lake Victoria, which bears the outlet of the White Nile. Kagera River contributed 9% to 10% of the total Nile Waters [10]. Hydrological measurements are essential for the interpretation of water quality data and for water resource management.

Variations in hydrological conditions have important effects on water quality. In rivers, such factors as the discharge (volume of water passing through a cross-section of the river in a unit of time), the velocity of flow, turbulence and depth will influence water quality [11]. The annual rainfall varies from 700 - 1400 mm in the East and lowlands of the West to 1200 - 1400 mm in the high altitude region. The temperature regime is more or less constant with a temperature regime of 16°C - 17°C for high altitude region, 18°C - 21°C for the central plateau region and 20°C - 24°C for the eastern plateau and lowlands of the West. The climate is of the temperate equatorial continental type (AW3) according to KOPPEN classification .There are two rainfall seasons with the longer south-easterly monsoon rain between February and May, and the shorter north-easterly from September to November. The runoff responds to the rainfall with a higher peak in May and a smaller peak in November [12].

The upper tributaries are generally steep but include flatter reaches where swamps have formed. The average discharge is estimated to be 256 m³/s with a low flow of 85 m³/s at Kagera. Says that the Kagera is the main affluent of the Lake Victoria with a middle debit of 256 m³/s and has this title considered as the source of the Nil. One will note that has the entry of the Kagera in the Victoria lake, the debit of the Kagera is of 262 m³/s. The main rivers in the Nile Basin of Rwanda are: Mwogo, Rukarara, Mukungwa, Base, Nyabarongo and the Akanyaru of which the water are drained by the Nyabarongo which becomes Kagera at the outlet of Rweru Lake [13].

Water Use and Water Demand Projection

The pressures on water resources primarily result from utilizing the natural resources to meet basic needs as well as social-economic development. The effects of water resources use are demonstrated in the changes in the quantity and quality of water. All aspects of human activities in Rwanda have produced varying impacts and degrees of modification to the available water resources and these impacts are manifest at the catchment and sub-catchment levels as the following examples illustrate.

4.1.1. Domestic Water Use and Projected Water Demand

The data from Energy and Water Sanitation Authority show that the water demand in Kigali city is 55,080 m³ per day, whereas nominal production is 30,525 m³ per day. This illustrates a deficit in the drinking water requirements for Kigali. It is estimated that water demand over the next decade will double in Kigali and rural areas and more than double for the semi-urban settlements. **Table 2** shows the estimated consumption of water in 2005 and the projected water demand for 2020 based on Status, Trends and Key Issues in Water Resources Demand and population growth [14].

4.1.2. Industrial Water and Agriculture Water Use

The study on the Knowledge and Management of Water data done under the preparation of the National Management of the Water Resources Project [15] indicates that industrial water requirements will be between 300,000 and 900,000 m³/yr by 2020 in urban areas. Rwanda's economy is based on agriculture, and agriculture is rain-fed and is therefore exposed to vagaries of climate fluctuation. Many areas which use poor farming methods without integrating soil and water resources conservation tend to have weak agricultural productivity. In such instances, soil moisture becomes the limiting factor for crop growth. In 2000, the total national water withdrawal for agriculture was estimated at 150 million·m³/year and the share of agriculture estimated at 68%. Rice growing (on 8500 ha) constituted the crop that used most irrigation at about 25,500,000 m³. The government is making efforts to promote irrigation including hillside irrigation, especially in the dry lands of the eastern province, to increase food security. This is in light of research that projects the agricultural water requirements for 2020. It shows that the eastern region will require more water than at present, while the Congo-Nile ridge region

Table 2. Potable water demand.

Types of Settlements	Estimated Consumption (l/home/day)	Projected Needs for 2020 (l/home/day)
Rural	10	20
Semi-Urban	35	90
Kigali	48	90

which will have enough rainfall will only require the development of rain harvesting systems to allow the utilization of water in drier seasons for food security purpose. **Figure 3** shows the agricultural water requirements in 2020 [16].

5. Cooperation and Regional Conflicts in River Basins

5.1. Cooperation and Water Conflict in the Literature

The international nature of many of the world's great rivers and increasing scarcity of water has led to discussions in the academic literature of the growing potential for violent conflict, or "water wars", between nations over shared water resources [17] [18]. A related body of literature links environmental scarcity in broader terms with conflict [19]. The potential for security problems or violent conflict at an international or sub-national level has also been discussed in relation to the impacts of climate change and, more recently, in a special issue of Political Geography [20].

5.2. Forms of Conflict and Cooperation

The development of Water Event Intensity Scale [21], which draws from the International Cooperation and Conflict Scale. The scale ranges from extreme conflict at –7, for a formal declaration of war, through to extreme cooperation at 7 for voluntary unification into one nation. In this scale, conflictive interactions include hostile verbal expressions (official or unofficial) and hostile diplomatic, economic or military acts. The same cases of conflicts also happen in Nile Basin.

5.3. Current Conflicts and Treaties on Nile River

The issue of Conflicts has never been far from the banks of the Nile. The same is true of most international waterways, but none to the degree of the Nile. The source in central Africa; its value to the 11 countries through which it flows and the total dependence of Egypt and Sudan on this life line have always made the political and biological life of the river a source of conflict, the Egyptians see the river as a gift from God to them. Without the Nile there would have been no Ancient Egypt and its great heritage, only desert. But Ethiopia, where most of the river's water originates, also wants to make use of it and has been planning a huge dam for years. On completion the Renaissance dam in 2017, costing $4.3 billion and spanning the Blue Nile at the border with Sudan, is to generate 6000 megawatts of electricity. Ethiopia began to change the course of the river, displacing it by several hundred meters, in a move that has outraged Egypt and generated near panic over future water supplies. Egypt depends on the Nile for 98 per cent of its water and water is in increasingly short supply. The Arab world's

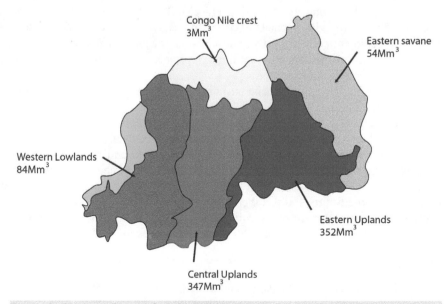

Figure 3. Water requirement in 2020 (million cubic meters) for agriculture in Rwanda.

most populous country has seen its population soar from 53 million in 1990 to more than 80 million today.

Egypt is determined to uphold an agreement dating back to 1929 and the colonial era. This document provides for Egypt and Sudan to have rights to more than 80 per cent of the water, even though the Nile flows through 11 countries in all. The other countries aim to change this provision. Ethiopia's parliament has now passed the ratification of a treaty intended to replace the old Nile Water Agreement, in terms of which a joint forum will decide on how the waters are used. But Egypt has rejected this, insisting on its prior rights, in the longer term, Egypt will have to make concessions to Addis Ababa as Ethiopia wants to solve this by diplomatic negotiations [22].

5.4. Benefits, Conditions and Limitations of Cooperation

International river basins cooperation is seen as desirable [23] describe four types of benefits: firstly, benefits to the river from cooperative basin-wide environmental management, such as improvements in water quality and maintenance of biodiversity; and secondly, benefits from the river, such as hydropower, irrigation, flood and drought management and navigation. Thirdly, they hypothesize benefits because of the river, for example reduced risk of conflict between riparian nations and increased food and energy security, and fourthly, benefits beyond the river, such as integration of regional infrastructure, markets and trade. Sadoff and Grey suggest that there are costs to non-cooperation, as well as to cooperation and that, depending on the particular circumstance, the scale of benefits may or may not outweigh the costs of cooperation. In the absence of strong cooperation, [24] assert that the varying intensities of conflict that commonly exist, but fall short of violent conflict, nevertheless have negative consequences on the less powerful riparian countries. It is clear that all riparian countries of Nile Basin should cooperate each other to avoid any kind of water war.

6. Conclusions

Even though Rwanda is well reported in water resources management progress; the country faces a number of problems in relation to the sustainable management and protection of these resources, which will ultimately, if not adequately and timely addressed, detrimentally affect the country's economic and social development. The provision of an enabling environment within which sustainable water resources management can be achieved includes the preparation of a comprehensive National Water resources management policy, the promulgation of legislation and appropriate regulations which will allow the implementation of the policy, and the establishment of a rational institutional framework through which all government levels can function. Therefore, the following issues need to be addressed:

1) Frequent assessment and monitoring of water resources including the collection, analysis, storage and dissemination of water related information; availability of reliable data is necessary to help governments in appropriate decision making.

2) Capacity building to enhance managerial and technical skills within water resources sector.

3) The protection of critical environmental functions to ensure the sustainability of the source of water resources, conflicts in the region affecting effective participation of countries and time implementation of projects.

4) Broadening participation at the country level-through projects and multiplier effect, and other activities including partnership for watershed management.

5) Increase country contribution and financial sustainability of the institution; strengthening Linkages of national and regional initiatives and programs.

6) Institutional arrangements at country level for improving data collection and management, government to invest in water and related resources monitoring and effectively share data.

However, to deal with challenges for water resource development and management in the Nil Basin there should be a transitional institutional mechanism for co-operation, an agreed vision and basin-wide framework, and a process to facilitate substantial investment in the Nile basin. It is based on the recognition that the basin has a shared past and a shared future, and that there is an urgent need for development and for the alleviation of poverty. It represents deep commitment by the Nile riparian countries to foster co-operation and pursue jointly the sustainable development and management of Nile water resources for the benefit of all.

Acknowledgements

This study is supported by the National Science Fund of China (No.41272249), Research Fund for the Doctoral Program of Higher Education of China (No.20110072110020).

Conflict of Interest

The authors declare that they have no conflict of interest.

References

[1] CGIS-UNR. Environment and Poverty (2003) A Pilot Study on the Relationship between Environment and Poverty Mapping. UNDP, Kigali.

[2] MINALOC (2003) The Draft of National Policy for the Management of Risks and Disasters in Rwanda.

[3] MINAGRI (1998) National Policy for the Management of Water Resources.

[4] NISR (2012) Rwanda Census. The National Institute of Statistics of Rwanda. Kigali, Rwanda.

[5] Nzeyimana, L. (2003) Rusumo Dam-Social Challenge in Kagera River Basin: Participation of the Affected Population. Linköping University, Sweden.

[6] Ndayiragije, D. (1992) A Geomorphological Study of the Kagera Basin. Unesco/University of Burundi, Bujumbura.

[7] Twagiramungu, F. (2006) Environmental Profile of Rwanda. National Authorizing Office of FED and European Commission.

[8] NELSAP (2001) 11-12 December Report from NELSAP River Basin Project Preparatory Workshop. Nile Basin Initiative.

[9] NBI (2001) Water Resources Planning and Management, Project Document. Nile Basin Initiative.

[10] NBI (2001) Transboundary Environmental Analysis. Nile Basin Initiative.

[11] Ndege (1996) Strain, Water Demand, and Supply Directions in the Most Stressed Water. Civil Engineering Department, University of Nairobi, Nairobi.

[12] NBI (2005) National Nile Basin Water Quality Monitoring Report for Rwanda. Nile Transboundary Environmental Action Project, Nile Basin Initiative.

[13] RNRA (2012) Water Resources Information Bulletin for June-July 2012. Rwanda National Resources Authority, Kigali.

[14] MINITERE (2005) Project of National Management of Water Resources. Components D: Technical Studies. The Ministry of Lands, Environment, Forestry, Water and Mines.

[15] PGNRE (2005) Knowledge and Data Management on Water in Rwanda. General Technical Report, Kigali.

[16] K V P. (2006) Analysis on the Integrated Management of Water Resources in Rwanda. Working Paper for the NGO Protos, Final Report.

[17] Toset, H.P.W., Gledisch, N.P. and Hegre, H. (2000) Shared Rivers and Interstate Conflict. *Political Geography*, **19**, 971-996. http://dx.doi.org/10.1016/S0962-6298(00)00038-X

[18] Gleditsch, N.P., Furlong, K., Hegre, H., Lacina, B. and Owen, T. (2006) Conflicts over Shared Rivers: Resource Scarcity or Fuzzy Boundaries? *Political Geography*, **25**, 361-382. http://dx.doi.org/10.1016/j.polgeo.2006.02.004

[19] Ronnfeldt, C.F. (1997) Three Generations of Environmental Research. *Journal of Peace Research*, **34**, 473-482. http://dx.doi.org/10.1177/0022343397034004009

[20] Nordås, R. and Gleditsch, N.P. (2007) Special Issue on Climate Change and Conflict. *Political Geography*, **26**, 627-736. http://dx.doi.org/10.1016/j.polgeo.2007.06.003

[21] Wolf, A.T., Yoffe, S.B. and Giordano, M. (2003) International Waters: Identifying Basins at Risk. *Water Policy*, **5**, 29-60.

[22] Dudin, M. and Frentzen, C. (2015) Egypt-Ethiopia Conflict over Nile Waters Flares. http://www.rappler.com/world/31553-egypt-ethiopia-conflict-over-nile-waters-flares

[23] Sadoff, C.W. and Grey, D. (2002) Beyond the Rivers: The Benefits of Cooperation on International Rivers. *Water Policy*, **4**, 389-403. http://dx.doi.org/10.1016/S1366-7017(02)00035-1

[24] Zeitoun, M. and Mirumachi, N. (2008) Transboundary Water Interaction: Reconsidering Conflict and Cooperation. *International Environmental Agreements*, **8**, 297-316. http://dx.doi.org/10.1007/s10784-008-9083-5

Water Pollution and Environmental Governance of the Tai and Chao Lake Basins in China in an International Perspective

Lei Qiu[1,2,3], Meine Pieter Van Dijk[3]*, Huimin Wang[1,2]

[1]State Key Laboratory of Hydrology-Water Resources and Hydraulic Engineering, Hohai University, Nanjing, China
[2]Research Institute of Management Science, Business School, Hohai University, Nanjing, China
[3]UNESCO-IHE Institute for Water Education, Delft, The Netherlands
Email: *mpvandijk@iss.nl

Abstract

The Tai and Chao Lake basins are currently facing a serious water pollution crisis associated with the absence of an effective environmental governance system. The water pollution and the water governance system of the two basins will be compared. The reasons for water pollution in both basins are similar, namely the weak current water environmental governance system cannot deal with the consequences of the rapidly growing economy. China's water governance system is a complicated combination of basin management with both departmental management and regional management. There is an absence of legal support and sound coordination mechanisms, resulting in fragmented management practices in the existing water environmental governance system. A comparison is made for the Tai and Chao Lake basins and Canada, France, the United Kingdom and the United States. Based on China's present central-local governance structure and departmental system, an integrated reform of basin level and water environmental governance in China should learn from international experiences. The reforms could consist of improved governance structures, rebuilding authoritative and powerful agencies for basin management, strengthening the organizational structure of the basin administrations, improving legislation and regulatory systems for basin management and enhancing public participation mechanisms.

Keywords

Water Pollution, Water Governance, Environmental Governance, Basin Management, Tai Lake

---

*Corresponding author.

Basin, Chao Lake Basin, Participation

1. Introduction

The five largest freshwater lakes in China—Poyang Lake, Dongting Lake, Tai Lake, Chao Lake, and Hongze Lake—are all located in the middle and lower reaches of Yangtze River and Huai River. The region has a high population density and has shown rapid economic development. These five lakes are major freshwater storage areas, and play an important role for maintaining a regional ecological balance, providing freshwater resources, preventing floods, diminishing environmental pollution, etc. However, with rapid industrialization and urbanization, serious environmental problems manifest themselves in these five freshwater lakes. Especially the Tai and Chao Lake areas have been listed as objects of Water Pollution Control and Management Projects by the State Council of China as long ago as 1996.

The Tai Lake basin is the core of Yangtze River Delta region, which is one of the most developed areas in China. Since the 1980s, the water quality level of Tai Lake basin on average decreased one level each decade [1]. A lot of water pollution prevention and control works in Tai Lake have been undertaken; however, the works still lag behind the economic and social development of the Tai Lake basin and water pollution has exceeded the environmental capacity. Water pollution prevention and control in the Tai Lake Basin has become the focus of the state environmental protection work with an important demonstration effect for other Chinese lakes.

The Chao Lake basin is located in the centre of the Anhui Province. Economic activities are less developed and the urbanization level is lower in this basin. Currently the most serious issue of Chao Lake is water pollution. At the end of the 11th Five-Year Plan for National Economic and Social Development (2006-2011), the water quality of Chao Lake was Class V, and the whole lake had a mild level of eutrophication.

There is a tendency in China to consider water issues as technical problems which can be solved by building dams or sluices. However, water problems are not just technical problems. Water governance is important to provide an integrated approach to the real water issues, which is necessary because of the interrelated nature and complexity of issues like water availability, water quality and flood problems and the need to manage and coordinate the processes for identifying, sharing and solving these problems in a multidisciplinary way and involving all stakeholders concerned [2]. Water governance is defined in this case as involving stakeholders to deal with water issues in an integrated way.

Increasing complexity and uncertainty in river basin systems have created land and water management problems, but the limited capacity of state institutions to deal effectively with such conditions suggests that the current system must be reformed and that a more powerful system-response capability founded on inter-organizational collaboration should be developed. Effective governmental and non-governmental collaboration would enable management problems to be dealt with in a more efficient, effective and equitable manner [3]. Countries should make the necessary efforts to obtain effective participation in the planning and decision-making process of water resources management involving users and public authorities [4].

According to Foerster [5] three central components of purposeful and adaptive institutional innovation for environmental water governance have to be addressed, namely developing processes to reach an ecologically sustainable allocation of water resources, using instruments to provide for and protect water and finally the development of management frameworks for rivers with environmental water regimes.

The challenge of making central requirements work effectively at local levels is a common problem for environmental governance throughout the world and a reason to put China's experience in an international perspective. China's problem is not the absence of environmental laws, but the challenge of making them work [6]. This paper will compare the water environmental pollution conditions and the governance systems of the Tai and Chao Lake Basin and learn from experiences elsewhere in the world. The objective is to analyze the water pollution and environmental governance problems. The situation in China will be compared with other countries, in particular Canada, France, the United Kingdom (UK) and the United States (US). The objective is to suggest governance policies for freshwater lakes, which are under pressure. We will argue that China emphasizes technical solutions for water management and does not pay sufficient attention to governance issues. The research question is how does China deal with lakes in governance terms and how does this compare to international practices such as more participatory water governance, and unified leadership of and well specified rules for different organizations involved?

2. General Situation in the Tai and Chao Lake Basin

2.1. Natural and Social Conditions

The Tai lake is a large shallow lake and part of the river network in the Yangtze River Delta plain. The shoreline around the lake has a total length of 405 km and an average depth of 1.89 m. The Tai Lake serves as a natural water reservoir, the most important water source for surrounding regions. The area of the Tai Lake basin is 33,700 km^2, and the water area is 2338 km^2 (2010). In terms of administrative divisions, the whole area of the Tai Lake basin is divided over three provinces and one city: Jiangsu Province (52.6%), Zhejiang Province (32.8%), Shanghai City (14.0%) and Anhui Province (0.6%).The Tai Lake basin is one of the most developed areas in China, and has produced 10.8% of China's total GDP with 0.4% of China's land area and 4.3% of China's population. Its per capita GDP was about 2.5 times China's average per capita GDP in 2010 [1].

The Chao Lake is located in the centre of Anhui Province, on the left bank of the lower reaches of the Yangtze River with a lake area of 760 km^2 and a total basin area of 13,500 km^2. The Chao Lake basin belongs completely to the administrative divisions of Anhui Province. 33 Rivers flow into the Chao Lake, and eight main tributaries flow into the Chao River and then flow into the Yangtze River through the Yuxi River in the west. The Chao Lake Basin has an average annual surface water content of 5.36 billion m^3 with an annual average of 3.49 billion m^3 flowing into it.

In 2010 the population in the Chao Lake basin was about 10.6 million, 5.7 million of which live in the urban areas. The urbanization rate is 54%. The Chao Lake basin is a pivot area of Anhui province. Due to a multiple operation structure of combined plantation, aquatic breeding and animal husbandry its agriculture is well developed. Its industry is mainly based on machinery, electronics, chemical industry, metallurgy, textile, food processing and building materials. The GDP reached about 335.9 billion RMB, while the GDP per capita is 32,000 RMB, lower than the national average level; the importance of the agricultural, industrial and service sectors is 17.4%, 47.0%, and 35.6% [7]. The major natural and economic indicators of the Tai and Chao Lake basins are summarized in **Table 1**.

2.2. Water Pollution Conditions

According to Surface Water Environment Quality Standard (GB3838-2002), the water quality of Tai Lake was Class I~II in the 1960s and Class II in the 1970s. At the end of the 1980s, the water quality further degraded from Class IV to Class V, and even to Below Class V (worse than Class V) in the period from the end of the 1990s to early 2003 [8]. With heavy pollution, particularly of the water in the bays, the Tai Lake is currently included in the list of the most seriously polluted rivers and lakes in China. A massive appearance of algae in May 2007 degraded the water quality at the drinking water intake spot for Wuxi City in the Meiliang Bay, causing a water shortage supply. The toxic-algae bloom in the Tai lake basin made headlines around the world at the end of 2014.

Rapid economic development quickly gave rise to an increase in pollution emissions. The industrial pollution in Tai Lake stems mainly from the textile printing and dyeing industry, the chemical raw materials and chemical products manufacturing, the food manufacturing, etc. A large number of industrial enterprises with low-technology and serious pollution have been transferred to rural areas where there is relatively weak supervision and it is even more difficult to control industrial pollution. Hence a lot of industrial pollution enters the Tai Lake through the rivers. The proportion of non-point source pollution is also increasing in the Tai Lake area. For

Table 1. Comparison of social and economic status between the Tai and Chao Lake basin (2010).

Basin	Socioeconomic status	Administrative region	Basin area/km^2	Population/million	GDP/billion RMB
The Tai Lake basin	State key economic zone	Includes Jiangsu, Zhejiang, Shanghai, and a little area of Anhui	33,700	34.8	3821
The Chao Lake basin	Local economic zone	Belongs entirely to Anhui Province	13,500	10.6	336

example, in the 1980s the proportion of total phosphorus (TP) increased from 78% to 87%; the proportion of total nitrogen (TN) from 60% to 78% and the proportion of Chemical Oxygen Demand (COD) from 50% to 63% [1].

Already in the 1980s, the pollution load in the Chao Lake basin exceeded the carrying capacity of the water sources, and a series of environmental problems arose in 2010. Data from 28 monitoring sections in the rivers around the lake showed that 17.9% of the section water quality was class IV, and 35.7% was class V. The data from twelve conventional sections for water quality monitoring in Chao Lake indicated that 50% of the water was Class-IV, V and inferior Class V [9]. The overall water quality of Chao Lake has improved from moderate to mild eutrophication after the environmental management of the 11th Five-Year Plan.

The waste water discharge of the Chao Lake basin amounted to 400 million tons; among of them, of which industrial effluent accounted for 13.4%, and urban sewage for 86.6%, the COD was 176,000 tons, emission of ammonia and nitrogen was 14,000 tons [7].

3. Governance Systems of the Tai and Chao Lake Basins

3.1. Pollution, the Background

Since China's reform and opening up in 1978, it has achieved rapid industrialization and urbanization in the middle and lower reaches of the Yangtze and the Huai Rivers. At the same time waste dumping and polluted water have increased continuously. Environmental management could not keep up with the pace of economic growth, which is one of the most important causes of water pollution and eutrophication of the Tai and Chao lakes.

Environmental management of the Tai Lake basin is a typical trans-administrative water pollution problem, which mainly involves two provinces and one city. The Tai Lake Basin Authority was set up by the Ministry of Water Resources in 1984 to perform water resources coordination and management in the Tai Lake Basin, but its capacity of supervision and coordination is limited. The Tai Lake Basin Authority is not a powerful basin environmental management institution but a public institution with limited functions. It cannot play a leading role in water resources protection and water pollution control.

Prior to August 2011, the Chao Lake basin covered three cities, including Chao, Hefei and Lu'an. Moreover, the water environmental management departments involved water conservancy, environmental protection, urban construction, fishing, transport, tourism and other things. There was no unified management organization or coordination mechanism for a long time and the various regions and departments all pursued their own interest, which led to disorderly management. External diseconomies of water pollution are amplified due to scattered management of administrative districts and departments.

Another cause of serious water pollution is the current system of farming and aquaculture, which results in non-point source pollution. The five largest freshwater lake basins are not only the pivotal production areas of grain, cotton and oil, but also important areas for poultry farming and fish breeding. The agricultural non-point source pollution was caused mainly by fertilizer use. Especially nitrogen and phosphorus nutrient losses are an important factor leading to eutrophication.

The existing agricultural production mode has aggravated the non-point pollution. The internationally recognized standards of average chemical fertilizer use is no more than 22,500 kg/km^2 per year, whilst the average annual chemical fertilizer use in China in 2011 was 43,430 kg/km^2 per year, which is 1.93 times the upper safety limit. When it rains, a lot of residual nitrogen is washed out through drainage network system into the lake. Furthermore the aquaculture surrounding the lake has also impaired the water quality, because the baits used by the fishers have become fertilizer for harmful algae, thus increasing eutrophication.

30% of the water pollution in the Chao Lake comes from industrial wastewater and domestic sewage, and 70% from non-point source pollution. The non-point source pollution is greater than the point source pollution, and the main pollution components are nitrogen, phosphorus, nutrients and aerobic organisms. The reason for this is that pollution control of industrial wastewater and domestic sewage have received much attention, but the agricultural non-point source pollution control has not received enough attention [10].

Finally, reclaiming land from marshes and building sluices has restricted ecological functions. The rapid growth in population of the lake area has required more land. A large-scale reclamation activity was started in the middle and lower reaches of the Yangtze and Huai rivers in the 1950s. The water areas decreased about 13,000 km^2, which amounted to 1.3 times the total area of the present five freshwater lakes. Also, it reduced the lake's water volume by 50 billion m³, which amounted to 1.2 times the total volume of the present five freshwater lakes [10]. The reclamation has caused a sharp shrinking of the lake area, reduction of the environmental capacity of the ecosystem functions and of the purification abilities, which has accelerated the lake's eutrophication.

In China water management often implies looking for technical solutions to deal with the problems. However, building sluices and dams discontinues the natural process of water exchange among the lakes and the large rivers, disrupting the cleansing circulation of fresh water and the development of biological diversity and they still need to be managed. Taking Chao Lake as an example, originally it was a semi-closed water body with many rivers flowing into the lake but only a single channel out to the Yangtze River. To meet the demands of urban water supply, shipping, agricultural irrigation, flood control and drought and so on, control projects have been initiated. For example, Yuxi Sluice is a large control sluice at the estuary of Chao Lake flowing into the Yangtze River, which was built at 1967 and has played an important role in flood control and irrigation. Since then the water volume exchanged between Chao Lake and the Yangtze River decreased by 88.2%. The lake almost lost its original natural transfer capacity, which led to serious sediment deposition and the accumulation of pollutants, and aggravated eutrophication.

3.2. Tai Lake Basin Governance System

According to the Water Law of The People's Republic of China issued in 1988, the basic management system of water resources is a system of unified management, which combines hierarchical management and horizontal departmental management. The Water administration department under the State Council shall be in charge of the unified administration of water resources throughout the country. In 2002 the Amendments of the Water Law of the PRC (the New Water Law) made it clear that the state will apply the water resources management system through combining basin management (interpreted as dealing with the water's natural characteristics) with regional administrative management (dealing with the socio-economic environment). This is for the first time a legal recognition for "integrated basin management".

The Regulation for Administration of the Tai Lake Basin was put in place on November 1st 2011 after it was accepted by theStanding Committee of the State Council of the People's Republic of China. The Regulation indicates that water resources management system is a "combination of basin management with departmental management and regional management" and that "the Nation is responsible for establishing coordination mechanisms and making overall arrangements of major issues in the Tai Lake basin management". According to these laws and regulations, the water environmental governance system in the Tai Lake basin includes four important institutions (**Figure 1**), which do not always collaborate.

3.2.1. The Ministry of Water Resources or the Department of Water Administration

Article 12 of the New Water Law stipulates: The department of water administration of the State Council is responsible for the integrated administration and supervision of the nationwide water resources. Article 5 of the Regulation for Administration of the Tai Lake Basin (2011) reads: The Tai Lake Basin Authority should perform the supervision and administration endowed by the laws, administrative regulations and the departments of water administration of the State Council under the leadership of the Ministry of Water Resources. Thus the Tai Lake Basin Authority is the resident agency of the Ministry of Water Resources.

Moreover, The Water Resources Protection Bureau of Tai Lake Basin performs the administrative responsibility for water resources protection and water pollution control. It is under the double jurisdictions of the Ministry of Water Resources and the Ministry of Environmental Protection, and its administrative position is one level below the Tai Lake Basin Authority.

3.2.2. The Department of Environmental Protection

Article 7 of the Environmental Protection Law states: the Department of Environmental Protection is mainly responsible for the integrated administration and supervision of environmental protection. Thus the Ministry of Environment Protection has been authorized to take care of water environmental management. The East China Environmental Supervision Centre performs the supervision and administration duties related to environmental protection in East China as the resident law enforcement and supervision agency of the Ministry of Environment Protection.

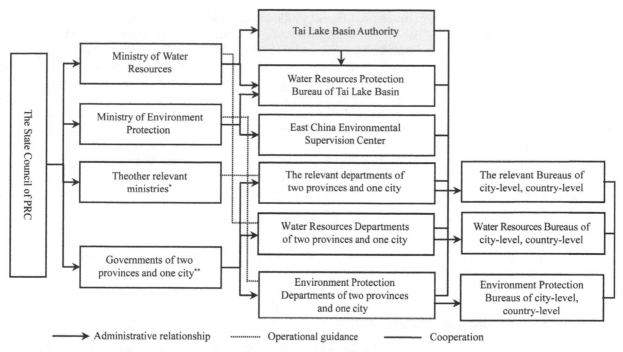

> → Administrative relationship ·········· Operational guidance —— Cooperation

*The other relevant ministries refer to agriculture, forestry, development and reform, fishery, tourism, and so on.
**Two provinces and one city refer to Jiangsu Province, Zhejiang Province, Shanghai.

Figure 1. Water environmental governance institutions of Tai Lake basin.

3.2.3. Other Relevant Ministries

Besides the Ministry of Water Resources and the Ministry of Environmental Protection, involved in managing of the water in the Tai Lake, the National Development and Reform Commission, the Ministry of Land and Resources, the Ministry of Agriculture, the Ministry of Forestry, the Ministry of Transport, the Ministry of Housing and Urban-Rural Development, and so on also perform these duties.

3.2.4. Local Governments and Relevant Departments of the Two Provinces

The relevant departments of local governments above the county level take charge of the administration of the Tai Lake basin in their own administrative regions according to the laws and administrative regulations. The responsibility and the evaluation system of local governments are focused on water resources protection and water pollution control in the Tai Lake basin.

Together these institutions have created a complicated governance system for the Tai Lake basin. Because the subjects of basin management and regional management belong to different institutional systems, the relation is not one of superior or subordinate but of equality as far as authority is concerned, which results in all kinds of difficulties in practice.

3.3. Chao Lake Basin System

Until August 2011 the Chao Lake basin had no unified management and no coordination mechanism existed. The environmental management of Chao Lake involved nearly ten departments, such as the department of environmental protection, water conservancy, development and reform, construction, agriculture, shipping and tourism, etc. Among them the departments of water conservancy, environmental protection and fishery carry out most of the work.

On August 22, 2011 the Anhui Provincial Government announced the abolishment of the Chao Prefecture-level Municipality administration structure and reallocated the districts and counties under its administration to other prefecture level municipalities. The administrative restructuring aimed to create a greater Hefei Municipality, which will foster development in the region, which will simplify integrated management of the Chao Lake basin.

The Chao Lake Administration Bureau was formally established on March 22012. According to the regulation in Opinions On Revocation of Chao City at Department Level and Adjustment of Part about Administrative Regions and Agencies Transfer and Staffing Placement [2011] 29 issued by the Anhui Provincial General Office, the Chao Lake Administration Bureau is an administrative institution of deputy department-level. It will manage the general affairs of Chao Lake as a whole, such as planning, water conservancy, environmental protection, fishery administration, navigation, tourism, and the major control facilities of the Chao Lake basin [11]. Since the establishment of the Chao Lake Administration Bureau is at its initial stage, there are some issues, such as the responsibilities of different levels of regional governments and the basin departments, which are not clear. Legislation and supporting regulation are lacking.

Similar to the governance system of the Tai Lake basin, the water environmental governance structure of the Chao Lake basin includes the department of water administration, the department of environmental protection, the other relevant ministries and the local governments and relevant departments of Anhui Province (**Figure 2**).

Major governance issues identified in the Chao Lake basin were scattered departmental management, and the need to establish coordination mechanisms among departments. Secondly, the pattern of regional segmentation, although improved in the Chao Lake basin, still requires further reforms. Finally, the basin management institution exists in name only, and the weak regional management capacity cannot achieve the management objectives. Carrying out the technical works is generally easy but managing the results is not.

3.4. The Chao Lake Governance System

The governance systems of the Tai and the Chao Lake basins represent the major water environmental governance mode of China. Water resources are state-owned, and the basin management system adopts a three-level structure of national ministries, basin administrations and regional departments. This is a combination of basin with departmental and regional management [12]. The lake basin governance system is slightly different from Lake Tai (cf. **Figure 1** and **Figure 2** and **Table 2** and **Table 3**).

3.4.1. National Ministries

The Ministry of Water Resources is responsible for most of the water resources management functions, but mainly focuses on water quantity, whilst the Ministry of Environmental Protection mainly focuses on water quality management, as the authority responsible for environmental protection. The Ministry of Agriculture,

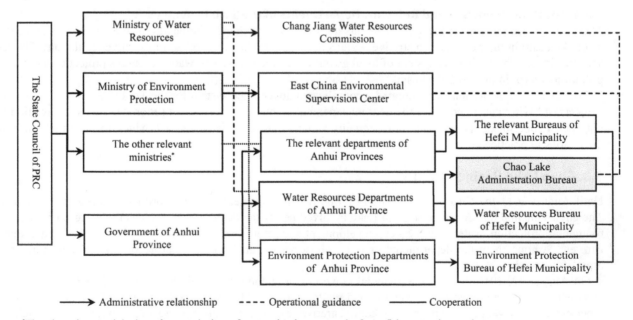

*The other relevant ministries refer to agriculture, forestry, development and reform, fishery, tourism, and so on.

Figure 2. Water environmental governance institutions of Chao Lake basin.

Table 2. Management system of water resources and environment.

Ministries	Main Functions
Ministry of Water Resources	Unified management of water resources, river management, flood control, water and soil conservation, management of hydraulic engineers
Ministry of Environment Protection	Water quality management, prevention and control of water pollution, water quality monitoring, water pollution source monitoring
National Development and Reform Commission	Water resources development, planning of ecological environment construction
Ministry of Agriculture	Non-point pollution control, protection water environment of fishery and aquatic wildlife habitats
Ministry of Forestry	Watershed ecology, protection and management of water conservation forests, wetland management
Ministry of Land and Resources	Land resources management, monitoring and supervising the excessive mining and pollution of groundwater
Ministry of Transport	River navigation, marine pollution control
Ministry of Housing and Urban-Rural Development	Project planning for urban water supply, drainage and sewage treatment

Table 3. Assessment of functions of water environment governance between the Tai and Chao Lake basin.

Functions		The Tai Lake basin	The Chao Lake basin
Decision making	Agencies	The Nation is responsible for making overall arrangements of major issues in the Tai Lake basin management, and the Ministry of Environment Protection works as the major authority of environment protection.	The government of Anhui Province and Chang Jiang Water Resources Commission are responsible for water resource management and water environmental protection of the Chao Lake basin.
	Legal guarantee	The Regulations for Administration of the Tai Lake Basin were put in place on November 1st 2011.	The Revised Regulations on Water Pollution Control of Chao Lake Basin were implemented on 1st December, 2014.
Execution	Implementation of policy and plan	Due to lack of effective enforcement means and efficient authority, the actual power of the Tai Lake Basin Authority to deal with water pollution is limited.	The three departments of water conservancy, environmental protection and fishery undertake major activities. It is difficult for the Chao Lake Administration Bureau to implement an effective management system.
	Enforcement Agency	No special enforcement agency	No special enforcement agency
Supervision	Enforcement capacity and power	It is to inspect comprehensive treatment for key areas and rivers, as well as regularly supervise performance of each member unit of the Provincial Committee of Tai Lake Pollution control.	Lack of law support and public participation
Coordination	Department coordination	The Nation is responsible for establishing coordination mechanism. Joint Conference at Province and Ministry Level for comprehensive treatment of Water environment in Tai Lake basin has been established.	There is no special department coordination mechanism.
	Regional coordination	Jiangsu Province, Zhejiang Province and Shanghai all set up their relevant coordination organizations.	The main problem is existing separate management of "the lake and rivers" of Chao Lake.

Ministry of Transport, National Development and Reform Commission and other departments respectively undertake the management functions of fishery resources, wetland, groundwater, etc. [13].

3.4.2. Basin Administrations

The Tai Lake Basin Authority performs basin management duties as the resident agency of The Ministry of Water Resources. Its duties on environmental protection are protecting water resources of the Tai Lake basin, partitioning water environmental function areas, examining water environmental carrying capacity, establishing a total pollution control plan, monitoring water quantity and water quality at the provincial boundaries, in important water areas and in case of inter-basin water transfers.

The Chao Lake belongs to the river network of the Yangtze River basin, so the Chang Jiang Water Resources Commission is responsible for water resource management and water environmental protection of the Chao Lake basin.

3.4.3. Regional Administrative Departments

Regional water resources departments and other relevant departments undertake and implement the specific functions of river basin management which are endowed by the central government. The departments have a presence at the province, city and county level. For the Tai Lake basin this includes the water resources, environmental, and other relevant departments of Jiangsu Province, Zhejiang Province and Shanghai. For the Chao Lake basin it includes the relevant departments of the Anhui Province and the Hefei Municipality. How the four functions work in the Tai and Chao Lake Basins is analysed in the following section, based on **Table 3**.

4. Governance Problems Analysis in China and International Experiences

4.1. Conflict of River Basin and Administrative Management

Although the Tai Lake basin has made some progress with integrated water resources governance since the 1980s, there is still a long way to go to improve the water governance system and coordination mechanisms. In the Chao Lake basin, the integrated water resources governance system is still at an early stage. The Chao Lake Administration Bureau was only set up in 2012, but it is not yet a lake basin management institution. Taking the examples of the Tai and Chao Lake basins, we will analyse water pollution and environmental governance problems.

As stipulated in Article 12 of the New Water Law, water resources management for river basin integrates regional administrative management in China. But river basin and regional management are different as concerns the units concerned and their objectives. The basin management regards the river basin as the unit to manage water resources, and its objective is to make overall effective utilization of the basin water resources. Regional management is based on the administrative district as the unit and its goal is to make comprehensive utilization of water resources to achieve regional economic development. The different administrative districts in the basin are more concerned with their own interests focusing only on economic growth, which causes constant conflicts between the local government and the basin management institution. The former represents the local economic interests and the latter represents the interests of the entire basin. Under the current management structure, the level of the basin management institution is significantly lower than the level of local government, whatever its authority or administrative power, so it is difficult for it to achieve the implementation functions of the basin management institution [14].

4.2. A Comparison with the Federal Environmental Protection Agency in the US

In the Chao Lake basin, the basin management institution depends strongly on the local government for the funds and personnel, so it is difficult to carry out its role. To the contrary environmental watershed management in the United States is carried out in a relatively centralized system under unified leadership. Between the federal and local agencies, power is mainly concentrated in the federal agency and among the federal agencies; power is concentrated in the federal environmental agency. Regional offices in various environmental protection areas are set up by the Federal Environmental Protection Agency (EPA) to exercise power on behalf of EPA. The EPA is responsible for formulating relevant regulation, providing specifications and standards as well as benchmarks, and supervising the implementation of other relevant institutions. The EPA has a special power, that is, it can replace other institutions and implement supervision and disciplinary measures for illegal activities under certain circumstances.

4.3. Absence of Powerful Execution Agencies in River Basin Management

For a long time, the basin administration agencies in China were responsible for the formulation of overall basin utilization plans and relevant professional plans, but power of authority for supervision and implementation of the plans is not clearly specified. Since no clear legal guarantee or responsibilities exist, the basin agency cannot receive recognition and support from other departments when performing its duties, so its roles are far from what they should be.

The Tai Lake Basin Authority was set up for example by The Ministry of Water Resources in 1984 to perform water resources coordination and management in the Tai Lake Basin, but its capacity of supervision and coordination is limited by its defined responsibilities and legal status. It is not a powerful basin environmental management institution but a public institution with administrative functions, and does not play a leading role in water resources protection and water pollution control. When dealing with an acute water pollution incident, the Ministry of Environment Protection will be in the forefront, while the basin management institution just implements policies without having decision-making authority. It only reports to its superior, the Ministry of Water Resources, which puts the basin management agency in a passive position.

4.4. A Comparison with the Thames Water Authority in the UK

International experience shows that a powerful river basin management institution with comprehensive decision-making and coordination mechanisms should be built for integrated basin administration [15]. For example, the Thames Water Authority was created in the United Kingdom in 1975. It brought together all the water management functions of the region in one public body to strengthen the macro-management of central government [16].

4.5. Poor Coordination of the Departments

The coordination of the departments is a big issue under the existing water environmental governance system. There are water related laws and regulations in water administrative departments, in environmental protection departments, and other departments such as agriculture, fishing as well as shipping. Considering their own interest, these departments make regulations on their own without a coordination mechanism, which leads to conflicting legislation. For example, the Department of Environment Protection made a Plan for Water Environmental Protection and a Plan for Prevention and Control of Water Pollution, while the Department of Water Resources made a Plan for Water Resource Protection, and the Department of Environment Protection issued the Water Environment Functional Zoning, while the departments of water resources issued the Water Functional Regionalization. Due to the conflict of interests and the lack of communication, the evaluation standards of water environment and water quality adopted by each department are different, so the water quality information and the amount of pollution emission released by each department are different, which seriously damages the government's credibility.

4.6. A Comparison with Coordination in Canada

Canada issued regulations and systems about the protection of ocean environment, which clarify the governance structure by specifying the responsibilities, status, rights of the relevant institutions, and the relationships with local governments. It attaches great importance to the comprehensive management patterns in legislation so as to avoid many problems.

4.7. Unsound Supervision Mechanism and Public Participation

Presently the supervision mechanism of China's water environmental governance system has many shortcomings. There is no special enforcement institution in the basin management system, nor does the law have clear rules allowing the system to discipline those who do not fulfil their duties. Especially the environmental protection supervision and management personnel tend to abuse their powers and to neglect their duties.

The public are the direct victims once the water environment is polluted, so the public deserves to participate in decision-making and supervision mechanisms. However, public participation is missing, and there are few ways and channels to obtain environmental information for the public, which leads to a low citizen's awareness of water quality degradation.

4.8. Public Participation Mechanisms in France and the US

More and more countries are actively introducing public participation mechanisms in water environment protection. In the United States, many feasible roads have been taken to encourage the public to participate at federal, state and local levels, including public hearings, large public meetings, public committees and so on. The government cannot make any major decisions without consulting public opinion. In France, the river basin commission is comprised of experts, voters and user representatives, each of whom account for a third of the "Three-Thirds" System. This management pattern not only considers the council's authority and professional inputs, but also fully meets the citizen's right to know, which greatly improves the efficiency of river basin environmental management.

5. Conclusions and Policy Suggestions

Comparing how other countries deal with polluted lakes it comes out that the emphasis in China is too much on looking for technical solutions, rather than improving the governance systems for these lakes. The trends in international water management are from regional to river basin management and from decentralized to unified management. Learning from the experiences of foreign management of water pollution control, the feasible reform of the current river basin management system in China is still basing the management system on the government hierarchy, which integrates departments and regions, and deals with the conflicts among different departments and regions in the river basin by means of rebuilding organizations, improving coordination mechanisms and more legislation. The international practices suggest unified leadership, creating an integrated basin management institution or defining more precisely the specified responsibilities of the institutions involved in water governance.

Based on the above discussion and analysis, we would suggest the following policy suggestions for the water environmental governance in China, especially for the Tai and Chao Lake basin.

5.1. Improve the Authority of Basin Management Agency

Water environmental management in the Tai Lake basin is a typical trans-administrative pollution problem. So the Tai Lake Basin Authority should be elevated to the level of resident agency of the State Council of the PRC higher than the administrative regions. It would then be responsible for formulating the laws and regulations of water environment governance in each administrative region, free from any interference by local governments.

In the Chao Lake basin, it is necessary to define the legal status of the Chao Lake Administration Bureau as the central Chao Lake management agency, which shouldbe responsible for pollution control, resource protection and ecological construction within Chao Lake Basin. It should exercise the relevant supervision and management and administrative enforcement duties related to water conservancy, environmental protection, land and resources, transportation, forestry, agriculture, and fisheries, etc.

5.2. Strengthen the Organization Structure of Basin Administrations

To achieve integrated basin management, we suggest an organization structure of basin administrations with decision-making, implementation, supervision, and coordination functions.

1) The basin decision-making and coordination agency will be the highest authority. The great and significant river basins which concern national sustainable development should be directly managed by the Central Government, such as the Tai Lake Basin. Its members should include the directors of governments and departments, the heads of the administrative regions in basin, and a number of water user representatives. Its main function would be to formulate policy and mid-long term basin plans, schemes of water resource utilization and pollution emission of administrative regions. Additionally, it should handle basin water disputes and environmental pollution, and coordinate the departments of environmental protection, water conservancy, fishery and construction and to provide information.

2) There should be a basin executive agency in charge of the routine work of basin management. Its members should consist of management staff and experts with practical experience. Its main function would be to be responsible for the specific water matters of the basin, including a regular release of information concerning the water quality of each monitoring section. The water resources management sectors and environment protection sectors of provinces, cities and counties in the basin are only responsible for carrying out the schemes of water resource utilization and pollution emission transmitted by the basin executive agency.

3) The Basin supervisory agency must be an independent body, with as its main function the monitoring and evaluation of the effects of the basin management institutions, implementing national laws and relevant policy, basin planning, etc.

5.3. Improve Laws and Regulations System of Basin Management

The government needs to strengthen water resources management laws. Although the department of water administration understands the importance of basin management from the successful experiences of foreign basin management, the New Water Law has chosen for a "combination of basin management with administrative region management", where there is no clear definition of Basin Management in the existing laws.

It therefore follows that we require a special law on Basin Management redefining the nature of the river basin administrative institutions, which should be independent of the administrative system. The law should endow the basin institution with an independent legal status and clear duties; expand the functions and powers of the river basin administrative institution to all aspects of the basin water pollution prevention and control; refine the duties of all levels of governments in water resources protection work and clarify the legal relationship and responsibilities between the river basin administrative institutions and government water administration departments.

5.4. Enhance Public Participation Mechanism

The following three communication rules are needed particularly to improve public participation in water environment administration. Firstly, respecting the public's right to know, establishing an effective information platform and releasing information timely, making the public understanding the water pollution situation, the plan and measures, etc. Basin management institutions should publicize the decision-making process and the relative information regularly.

Secondly, establishing good communication channels, forming a basin management committee composed of the public, the governments and the water management departments, asking the public to participate in decision-making, to reflect their own opinions with a desire to protect their own rights and interests, etc.

Thirdly, pay more attention to the functions of public supervision and to public opinions on environmental policy implementation.

Acknowledgements

This project is supported by the Major project of National Social Sciences Fund of China (Grant No.12&ZD214). In particular, we are thankful to the Tai Lake Basin Authority and the Chao Lake Administration Bureau for providing us with essential information.

References

[1]   Gao, J. and Jiang, Z. (2012) Conservation and Development of China's Five Largest Freshwater Lakes. Science Press, Beijing.

[2]   Van Dijk, M.P. (2012) Introduction. *International Journal of Water*, **6**, 137-154. http://dx.doi.org/10.1504/IJW.2012.049493

[3]   Watson, N. (2004) Integrated River Basin Management: A Case for Collaboration. *International Journal of River Basin Management*, **2**, 243-257. http://dx.doi.org/10.1080/15715124.2004.9635235

[4]   Salman, S.M.A. and Bradlow, D. (2006) Regulatory Frameworks for Water Resources Management: A Comparative Study. The World Bank, Washington DC. http://dx.doi.org/10.1596/978-0-8213-6519-9

[5]   Foerster, A. (2011) Developing Purposeful and Adaptive Institutions for Effective Environmental Water Governance. *Water Resource Management*, **25**, 4005-4018. http://dx.doi.org/10.1007/s11269-011-9879-x

[6]   Guttman, D. and Song, Y. (2007) Making Central-Local Relations Work: Comparing America and China Environmental Governance Systems. *Frontiers of Environmental Science Engineering in China*, **1**, 418-433. http://dx.doi.org/10.1007/s11783-007-0068-3

[7]   Ge, C. and Cao, Y. (2012) Institutional Study on Chao Lake Integrated Water Environmental Management. EARD, Asia Development Bank.

[8]   Jiao, C. (2010) The Present State and Challenges of Watershed Governance in China: A Study of Cases in the Taihu Lake Basin. *Journal of Kobe University Law*, **39**, 325-358.

[9]   Nie, J. and Chen, H. (2012) Opportunities and Challenges of Chao Lake Water Pollution Control after the Administrative Division Adjustment. *Chinese Journal of Environmental Management*, **4**, 35-38.

[10]   Jin, G. and Gao, C. (2008) Ecological and Environmental Problems of the Chao Lake and Countermeasures. *Journal of Yangtze River*, **39**, 80-82.

[11]   Zhang, H. and Cao, S. (2012) Anhui Province Performs Governance Strategy for Chao Lake. *Environmental Protection*, **18**, 59-61.

[12]   Van Dijk, M.P. and Liang, X. (2012) Beijing, Managing Water for the Eco City of the Future. *International Journal of Water*, **6**, 270-289. http://dx.doi.org/10.1504/IJW.2012.049500

[13]   Xia, J., Liu, X. and Li, H. (2009) Comparison of the Basin Management between the Murray-Darling Basin and Huaihe River Basin. *Resource Science*, **31**, 1454-1460.

[14]   Qian, Y. and Liu, Y. (2010) Study on Watershed Environmental Governance in China. *Ecological Economy*, **1**, 162-165.

[15]   Shen, D., Wang, H. and Jiang, Y. (2004) Riverbasin Management Organization: An International Comparison and Suggestion on China. *Journal of Natural Resources*, **19**, 86-95.

[16]   Shi, H. (2009) Comparative Analysis of Water Pollution Governance of Thames River Basin and Taihu Basin. *Water Resources Protection*, **25**, 90-97.

Legislation

Water Law of The People's Republic of China 1988 http://www.chinawater.net.cn/law/WaterLaw.htm
    Amendments of Water Law of The People's Republic of China (New Water Law) 2002
http://www.chinawater.net.cn/law/WaterLaw.htm
    Environmental Protection Law of The People's Republic of China 1989
http://www.chinabaike.com/law/zy/xf/cw/1331164.html
    The Regulations for Administration of the Tai Lake Basin 2011
http://law.lawtime.cn/d684161689255_1_p1.html
    Opinions on Revocation of Chao City at Department Level and Adjustment of Part about Administrative Regions and Agencies Transfer and Staffing Placement 2011

8

Turkey Creek—A Case Study of Ecohydrology and Integrated Watershed Management in the Low-Gradient Atlantic Coastal Plain, USA

Devendra Amatya[1], Timothy Callahan[2], William Hansen[3], Carl Trettin[1], Artur Radecki-Pawlik[4], Patrick Meire[5]

[1]Center for Forested Wetlands Research, USDA Forest Service, Charleston, USA
[2]Department of Geology and Environmental Geosciences, College of Charleston, Charleston, USA
[3]Formerly with Francis Marion and Sumter National Forests, USDA Forest Service, Columbia, USA
[4]Agricultural University of Krakow, Krakow, Poland
[5]Department of Biology, University of Antwerp, Antwerp, Belgium
Email: damatya@fs.fed.us, callahant@cofc.edu, williamfhansen@fs.fed.us, ctrettin@fs.fed.us, rmradeck@cyf-kr.edu.pl, patrick.meire@uantwerpen.be

Abstract

Water yield, water supply and quality, wildlife habitat, and ecosystem productivity and services are important societal concerns for natural resource management in the 21st century. Watershed-scale ecohydrologic studies can provide needed context for addressing complex spatial and temporal dynamics of these functions and services. This study was conducted on the 5240 ha Turkey Creek watershed (WS 78) draining a 3rd order stream on the Santee Experimental Forest within the South Carolina Atlantic Coastal Plain, USA. The study objectives were to present the hydrologic characteristics of this relatively undisturbed, except by a hurricane (Hugo, 1989), forested watershed and to discuss key elements for watershed management, including water resource assessment (WRM), modeling integrated water resources management, environmental assessment, land use planning, social impact assessment, and information management. Runoff coefficients, flow duration curves, flood and low flow frequency curves, surface and ground water yields were assessed as elements of the WRM. Results from the last 10 years of interdisciplinary studies have also advanced the understanding of coastal ecohydrologic characteristics and processes, water balance, and their modeling including the need of high resolution LiDAR data. For example, surface water dynamics were shown to be regulated primarily by the water table, dependent upon precipitation and evapotranspiration (ET). Analysis of pre- and post-Hugo streamflow data showed somewhat lower but insignificant (α = 0.05) mean annual flow but increased frequency of larger flows for the post-Hugo compared with the pre-Hugo level. However, there was no significant difference in mean annual ET, potentially indicating the resiliency of this coastal forest. Although the information from this study may be useful for comparison of coastal ecohydrologic issues, it is becoming increasingly clear that multi-site studies may be warranted to understand these complex systems in the face of climate change, sea level rise, and increasing development in coastal regions.

Keywords

Francis Marion National Forest, Integrated Water Resource Assessment, Water Balance, Water Quality, Eco-Hydrologic Models

1. Introduction

Managing forested wetland landscapes to sustain water quality, water quantity, and productivity require a detailed understanding of functional linkages between ecohydrologic processes and management practices. Watershed-level hydrologic and nutrient cycling processes are complicated by the presence of varied land features such as forests, wetlands, riparian areas, uplands and water bodies, along with multiple land uses. Despite the importance of forested wetlands, many of them have been converted in the last five decades primarily as a result of agriculture and urbanization [1]. However, hydrologic modification started with colonization and rice culture, with many of these changes still present [2]. The land use change of wetlands and streams influences hydrologic processes, as well as water quality, because wetlands affect the runoff and the transport and exchange of materials between terrestrial and aquatic ecosystems. Understanding those impacts on material storage, interception, transport and transformation are needed to manage water quality and water resources [3] [4].

The forested landscape of the lower southeastern Coastal Plain is characterized by a series of marine terraces bounded by scarps defined by a series of complex material deposits with marine, estuarine and riverine erosion affecting their composition and surface and subsurface hydrology [5]. These landscapes consist of natural and managed forests, depressional wetlands, pine flatwoods, riparian areas with bottomland hardwoods (BLH), extending to maritime forests bounded by brackish to salt marsh [6]. They are generally characterized by poorly to moderately drained high water table soils, wet sites, and low topographic relief [7]. Some of these lands have been drained artificially to lower water tables for accessibility and reduction of excessive moisture for increased crop growth [8]. Roads, ruts, and firelines also modify surface and subsurface hydrology [9]. Hillslope processes dominate the hydrology of upland watersheds, but hydrologic processes on relatively low-gradient poorly drained Coastal Plain sites are usually dominated by shallow water table positions. In fact, most of the outflows from these watersheds drain from the relatively flat marine terraces into saturated riparian areas where the water is at the surface or a shallow water table is present, meaning that total outflow depends on the frequency and duration of flooding and on the dynamics of the water table [10]-[13]. All of these processes complicate the task of quantifying the water budget components of these forested wetlands, and the need to account for the use of various water and silvicultural management practices, and interactions with surrounding uplands makes the job even more challenging.

Some of the past studies have documented hydrology and water budgets for these forested wetlands [8] [10] [12] [14]-[26]. However, most of those studies were focused on a specific issue and/or were very site specific on a field-scale basis using mostly headwater watersheds, limiting the ability to respond to broader scale, more complex and integrated issues, including applications of social sciences and needs, most of which occur or exist at the larger watershed or regional scale. Such issues or problems are most efficiently addressed by using a multi-collaborative approach [7].

Watershed studies provide the needed context to integrate the spatial and temporal complexities that arise when assessing the interactions of land management practices across a complex physical and biological landscape. Accordingly, Turkey Creek watershed (WS 78) was instrumented for gauging in late 1963 and monitored until 1984 [27] on the Santee Experimental Forest to provide the basis for eco-hydrological monitoring and modeling studies within this predominantly forested 3rd order coastal watershed within the Francis-Marion National Forest in South Carolina, USA (**Figure 1**) [26]. This region is characterized by its rapidly growing population and associated residential and commercial development, and a forest resource base that supports both commercial values in terms of the wood products industry and societal values (e.g., example water supply, cultural history, scenery, and recreational activities). The most effective way to address these issues is through multi-cooperation among various disciplines and agencies for conducting long-term monitoring and modeling studies [7]. Therefore, recognizing the importance of long-term data from a relatively undisturbed large forested landscape in a rapidly changing coastal environment, the hydrologic monitoring was reactivated in 2005 by United States Department of Agriculture (USDA) Forest Service (http://www.srs.fs.usda.gov/charleston/) by installing a flow gauging station approximately 200 m upstream of the previous gauging station in collaboration with the College of Charleston and the United States Geological Survey (USGS).

The goal of this multi-collaborative approach was to transfer sound science-based information and provide a basis for continuing and expanded interdisciplinary research to address critical issues surrounding the sustainable management of present and future water resources on the low-gradient forested landscape of the southeastern Coastal Plain region, as was synthesized recently by [7] for research studies conducted in last 10 years. The

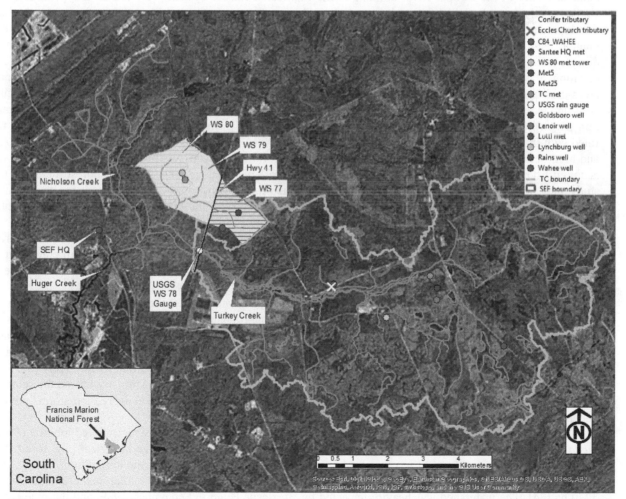

Figure 1. Location map of Turkey Creek watershed (WS 78) in green boundary using high resolution LiDAR data in 2011 (**Table 1**). Blue lines are streams and wetlands based on National Hydrography Data (NHD). Locations of weather stations, stream gauges, and ground water wells are also shown including for three other adjacent 1st (WS77 and WS80) and 2nd (WS 79) order watersheds within Santee Experimental Forest shown in red boundary.

main objective of this paper is to document the ecohydrologic characteristics and identify and discuss elements of watershed management including water resources knowledge and assessment and modeling integrated water resources management by providing science-based applications and tools, administrative policy and decision process, and socio-economics as related to water resources management on this pilot watershed.

The scale of the entire Turkey Creek watershed (not shown) with the 6th level 12-digit Hydrologic Unit Code (HUC) of 40 - 160 km^2 drainage area is an intermediate one for analysis of watershed conditions but large enough to have many of the hydrologic complexities suggested and be addressed at national forest planning scales, but small enough to be effectively evaluated, managed and addressed at larger project and watershed improvement scales [28] [29].

2. Watershed Description

As currently defined the 3rd order (1:24,000 scale) system at the gauging station, "Turkey Creek" (WS 78) is located within the USGS quadrangle maps of Huger (NE), Bethera (SE), Shulerville (SW and SE), and Ocean Bay (NW and NE) at 33°08'N latitude and 79°47'W longitude (approximate coordinate ranges of 610,400 to 628,600 easting and 3,658,500 to 3,670,500 northing [30]) in Berkeley County about 60 km northeast of the City of Charleston, South Carolina (SC), USA (**Figure 1**). Located within a 4th level 8-digit hydrologic unit code (HUC 03050201) of the Cooper River sub-basin [28], watershed WS 78 is at the headwaters of East Cooper River, a major tributary of the Cooper River draining to the Charleston Harbor System [31]. WS 78 is typical of other similar relatively undisturbed forested watersheds in the South Atlantic Coastal Plain that is facing rapid urban development. The drainage area of the watershed in this low-gradient landscape apparently evolved within the range of 3240 ha in 1964 when it was identified for the study to 7260 ha [15] [18] [30] [32]. Even with considerable attention, hydrologic boundaries within coastal areas may need review [33]. Most recently after the acquisition of high resolution LiDAR data its recomputed drainage area with consideration of boundary road cross-drainage culverts based on extensive field survey stands at 5240 ha [34]. Without considering those culverts and extent of ditching, the calculated area was 5985 ha as posted by the USGS at its gauging site information page (http://waterdata.usgs.gov/sc/nwis/uv?site_no=02172035). Much of these data have been further refined with improved stream gauging, soil, ecological and surface detail maps with increasing use of the LiDAR and high definition aerial photography in their development for more detail and precision. For example, newly acquired LiDAR data is being used to enhance the ecohydrologic mapping of the National Forest including the study site. Application of the absorption of the LiDAR pulses in water provides linear context for determining likely stream pathways and water extent that is further checked with the high resolution aerial photography (NAIP 2013 Imagery). LiDAR derived DEMs and shaded relief of added detail are helpful in defining watershed boundaries and linear features such as drainage ditches, old roads, dikes and water diversions [33]. Similarly, [35] divided groundwater aquifers into more units providing different naming conventions to them compared to the older documents. On the gauging side, two more continuous flow monitoring stations have recently been added in east (Eccles Road) and west (Conifer Road) tributaries of the study watershed (**Figure 1**).

The topographic elevation of the watershed varies from approximately 2 m at the stream gauging station to 14 m above mean sea level (amsl). The humid, sub-tropical climate is characteristic of the coastal plain having well distributed rainfall, hot and humid summers, moderate winter seasons, and exposure to coastal wind to hurricane events. Accordingly, the average annual air temperature and precipitation, based on a 63-year (1946-2008) record at the adjacent Santee Experimental Forest was recorded as 18.3°C and 1370 mm, respectively [3]. Seasonally, the winter is generally wet with low-intensity, long-duration (multiple-hour) rain events and the summer is characterized by high-intensity, short-duration (<1 - 4 hour) storm events; tropical depression storms are not uncommon [17].

Land use within the watershed is comprised of 44% (2306 ha) pine forest (mostly regenerated loblolly (*Pinus taeda* L.) and longleaf pine (*Pinus palustris*)), 35% (1834 ha) in thinned loblolly and longleaf pine forest, 8% (419 ha) mixed forest, 10% (524 ha) forested wetlands and water, and 3% (157 ha) in agricultural lands, roads and open areas. The calculated areas are based on percentage data [30] [32] and the total new drainage area of 5240 ha [34]. The watershed was heavily impacted by Hurricane Hugo in September, 1989, and the pine forest canopy was almost completely destroyed [36] while bottomland forests were relatively less impacted in the Category 4 winds. Most of the current forest stands on the watershed are approximately 25 years old with a mixture of remnant of large trees and natural regeneration.

2.1. Water Demand Management

There are not any publicized issues of water demand management (e.g. improved efficiency of water usage, recycling/reuse of water, improved efficiency of water supply etc.), given that in this watershed the few houses and the small agricultural areas depend upon groundwater. However, there remains substantial uncertainty relative to the effects of the past natural and anthropogenic drainage modifications and their influence on hydroperiod, flow permanence, and even the extent of past activities relative to watershed boundaries. We know that drought is a contributing factor, but we do not fully understand why over the past decade, a watershed with 52 km^2 goes dry about 30% of the time when it should be able to maintain perennial flow.

2.2. Water Resource Assessment on Regional Level

The regional water resources assessment looks both at surface and groundwater, including identification of the pertinent parameters of the hydrological cycle. The South Carolina State Water Assessment provides an overview of the State's surface and groundwater resources and includes information on the quantity, quality, availability and use of water in the State [37]. The assessment also describes existing water laws and regulatory programs, factors affecting surface-water availability and quality, and the State's ground-water distribution, well yields and chemistry. Conflicts in water availability stemming from the droughts of 1998-2002 and 2007-2008 are also summarized. The South Carolina Department of Natural Resources (SC DNR) is in the process of developing regional water plans for each of the State's eight major river basins that will be integrated into a composite state water plan (S. Harder, personal communication, 2015). This work includes the development of surface-water assessment models that will be used to evaluate the current and future availability of water in each basin as well as to test new water management strategies. The surface-water models are scheduled to be completed by the fall of 2016. Groundwater wise, regional groundwater levels have been historically reduced due to past withdrawals within the coastal areas. The withdrawals have lowered the hydrologic head that forces groundwater pressure and rate of flow within the aquifer, to streams and also riparian areas and wetlands. Anthropogenic "water uses" have also reduced levels of stored groundwater causing a predicted decline in streamflow (baseflow) [35] [38].

Detailed information on a regional assessment of the water resources of the South Carolina Coastal Plain and the Cooper River Basin where the study watershed (WS78) lies in the headwaters of East Branch of the Cooper River can be found in the water assessment report by the South Carolina Department of Natural Resources [37].

3. Water Resources Knowledge Base

3.1. Physical Data

The Lower Coastal Plain, where the study site is located, has a depositional topography caused by transgressions of the Atlantic Ocean with Pleistocene-aged sediments comprising the oldest exposures [39]. The regional remnants of sea level rise and fall are a series of beach terraces and back-barrier lagoonal sequences that roughly parallel the modern coastline. The geologically young topography of the Lower Coastal Plain is signified by shallow river valleys for streams originating within the Coastal Plain. We infer that in past times of lower sea level, these stream systems may have been more significant in their erosive power, which is relatively low at present.

Turkey Creek lies within the Talbot Marine terrace (also called Ten Mile Hill beds in some literature) which trends northeast to southwest, bounded on the northwest by the Bethera scarp and southeast by the more recent Suffolk scarp [5]. These landforms are often substantial determinants of hydrologic boundaries, but the marine and riverine erosion, as well as, in some instances, anthropogenic actions can modify hydrologic pattern and can alter coastal plain hydrology. **Figure 2** shows a break in the Suffolk scarp (brown to yellow) that resulted in drainage (light blue) from the Talbot terrace (Turkey Creek headwater vicinity in green and within the Cooper River system) into the somewhat lower Pamlico terrace and headwaters of Wambaw Creek (blue) that ultimately drains northeast into the Santee River system [33].

Through analysis of an online map [40], the dominant soil series on the watershed are poorly drained Wahee (clayey, mixed, thermic *Aeric Ochraquults*), and Lenoir (clayey, mixed, thermic *Aeric Paleaquults*) series [32] [41]. The watershed also contains small areas of somewhat poorly and moderately well drained sandy and loamy soils such as Lynchburg (thermic *Aeric Paleaquults*), Goldsboro (thermic *Aquic Paleudults*), and Rains (thermic

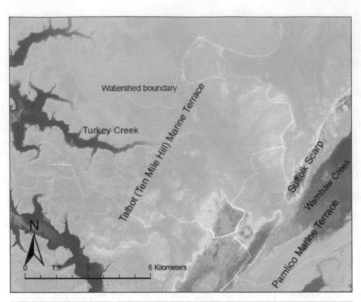

Figure 2. LiDAR DEM shaded relief image (1 - 16 m) showing Talbot and Pamlico Marine Terraces with break in Suffolk Scarp and 2005 USGS WBD estimated boundary which separated Turkey Creek watershed which contains WS78 gauging station in **Figure 1** and Wambaw Creek watershed.

Typic Paleaquults). Soils in the streambed and riparian buffers are comprised of the Meggett series (thermic *Typic Albaqualfs*). The right bank or northern half of the watershed contains sandy clay loams and silt loams, while the left bank or southern half is dominated by sandy clay loams. The permeability of these soils ranges from slow to rapid [40]. Overall, these soils have low infiltration capacity and high surface runoff.

3.2. Hydrological Data

Annual precipitation recorded during 1946-2008 period at the adjacent Santee Experimental Forest varied between 835 mm in 1954 to and 2026 mm in 1994 with an annual average of 1370 mm [3]. The minimum and maximum temperatures observed within the same period were −8.5°C and 37.7°C, respectively, with a daily average of 18.3°C. The 63-year (1946-2008) annual potential evapotranspiration (PET) estimated using the Hargreaves-Samani method [42] adjusted with monthly factors based on Penman-Monteith [43] ranged between 970 mm to 1304 mm, with an average of 1136 mm [3]. Thus the site has generally excess moisture as the average annual rainfall exceeds the average PET. However, records of hydro-meteorologic measurements (precipitation and streamflow) on the WS 78 exist only from 1964 to 1981 and again from 2005 onwards as described earlier. Studies conducted using the pre-Hugo data for (1964-1976) and post-Hugo data (2005-2014) are summarized below.

In a comparison of 5-year pre-Hugo daily flows of this study watershed (WS 78) with that from the adjacent 1st (WS 80) and 2nd (WS 79) order watersheds (**Figure 1**), it was hypothesized [44] that somewhat higher annual water yields from the largest WS 78 was possibly due to a difference in their land use, soils, and topography, increased base flows as well as potential discrepancies in drainage area calculation for flow estimate of WS 78 [34]. Using 13 years (1964-1976) of data, a large seasonal variability in storm event runoff coefficient (ROC) with higher values for the wet period than for the dry was found [18], potentially due to differences in forest ET that affected seasonal soil moisture conditions. This seasonal variability and evidence of ET influence has been further documented with more recent data sets from the first-order watershed WS 80 [12] demonstrating consistent results between studies. A recent hydrologic study [15] showed that the regenerated forest stand on the study watershed may be resilient to the natural disturbance such as Hugo in 1989 as indicated by the limited 4-year data of the measured stream flow response, compared to their expected values had the hurricane not damaged the forest. The spatially distributed hydrologic model Soil and Water Assessment Tool (SWAT) [45] was calibrated and validated to simulate daily and annual stream flow on this watershed [30], and that model is now being used to assess the future impacts of climate change on this landscape using scenario analyses.

Figure 3 shows the comparison of daily flow duration curves for a 10-year (1965-1974) period before Hurricane Hugo (1989) and a recent 10-year (2005-2014) period since regeneration of forest stands 15 years after Hugo. The smaller plot in the upper right hand corner is the blown up for 1% to 100% time a given flow exceeds. Data showed that the daily flow occurred about 73% of the time for the pre-Hugo period versus 67.5% of the time in the post-Hugo period. Clearly, flows of 2.5 mm day^{-1} or higher (occurring < 8% of the time) occurred more frequently in the post-Hugo period than the pre-Hugo, with the highest daily flow of 91.2 mm on October 24, 2008 as a result of over 150 mm rain in 24 hours in already wet antecedent condition. This trend is consistent with other regional studies reporting potentially higher more intensive storms and peak flows as a result of changing climate [3] [46]. The frequency of smaller flows below 2.5 mm day^{-1} (occurring more than 91% of time) was lower for the post-Hugo than for the pre-Hugo period, resulting in its lower total and average flow as will be shown below, potentially due to increased ET of the dense longleaf and loblolly pine stands regenerated after Hugo in 1989.

The peak flow rates for 100-, 50-, 25-, 10-, and 5-year return periods were computed by using Pearson III-type distribution in the flood frequency analysis with 13 years (1964-76) of pre-Hugo peak flows are shown in **Table 1** [44]. These results seem to be in reasonable agreement within the standard errors, except for the 50- and 100-yr periods with underpredictions as much as by 29%, compared to the results obtained for this study watershed using the USGS regional formulae [47] for rural coastal basins developed using data from South Carolina, North Carolina, and Georgia because of its location in rural areas of the South Carolina Coastal Plain.

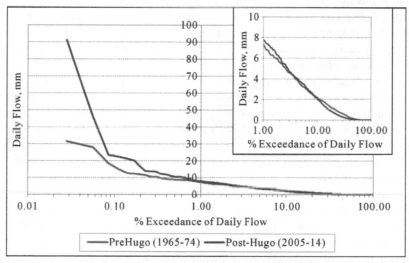

Figure 3. Measured daily flow duration curves for the 1965-1974 (pre-Hugo) and 2005-2014 (post-Hugo) periods for the study watershed.

Table 1. Estimated peak discharges (Q) and maximum discharges [47] predicted by Pearson Type-III (with a standard error in parentheses) and minimum discharges predicted by Gumbel distribution for various years of return periods for Turkey Creek watershed (WS 78).

Return period (year)	2	5	10	25	50	100	500
Peak flow formula [47]	$Q_2 =$ $60.3A^{0.649}$	$Q_5 =$ $123A^{0.627}$	$Q_{10} =$ $174A^{0.617}$	$Q_{25} =$ $245A^{0.606}$	$Q_{50} =$ $309A^{0.600}$	$Q_{100} =$ $380A^{0.594}$	$Q_{500} =$ $550A^{0.583}$
Q (m^{3}s^{-1})	12.0	22.9	31.5	42.9	53.2	64.2	76.9
Maximum discharge m^{3}s^{-1} by Pearson type-III (1965-1976)	15.9 (±3.2)	25.1 (±3.9)	30.6 (±5.1)	37.0 (±7.4)	41.5 (±9.5)	45.7 (±11.8)	
Minimum discharge m^{3}s^{-1} by Gumbel distribution	0.035	0.031	0.026	0.021	0.017	0.010	

However, it is important to acknowledge some uncertainty in this comparison because the estimates are from regionalized regression equations [47] with estimated error of 34% - 47.7%, depending on the percent chance exceedance event, and also did not use the peaks flows from this Turkey Creek watershed. Although both the methods used the Pearson III-type distribution, one reason for higher predictions by the USGS method was due to use of data from some rural basins with less than 10% urban areas in their regression models and also possibly due to the use of a generalized regional skew factor of −0.019. The results, except for the 40-yr and 100-yr periods, are comparable given the above reported range of average error predictions [47]. The limitation in estimates from the flood frequency formula should also be acknowledged as shown by [48] whose analysis shows a potential of 18% and 48% standard errors of estimates of mean annual discharge and a flood of 50-yr return period, respectively for this the 13-year period analyzed here.

The 100- and 50-year peak flow rates of 58.0, and 48.1 $m^3 \cdot s^{-1}$, respectively, estimated by the South Carolina Department of Transportation (C. Bodiford, personal communication, 2004) for their design of the highway bridge at the Turkey Creek gauge were 27% and 13.7%, respectively, higher on the conservative side than the Pearson III-type results (**Table 1**). At the same time the Gumbel distribution—also referred to as Fisher-Tippett Type I—was used to predict minimum discharges for 2- to 100-yr return periods for the Turkey Creek watershed in **Table 1** [44]. The predicted maximum and minimum flow values varied within about three-fold for 2- to 100-yr period. Although these high and low flow frequency discharge information are useful for stormwater management planning and design including environmental flow requirements for ecosystem restoration [49], the predictions developed using only 13 years of historic data need to be cautiously interpreted and re-evaluated using longer periods of observed data for more accurate predictions. Furthermore, although the study watershed went dry (zero flows as measured by the deployed sensor) about 1/3 of the time (mostly summer) for the recent post-Hugo period (**Figure 3**), very small flows difficult to be detected by the sensor may still possibly be occurring in the hyporheic zone of the stream bed.

The annual rainfall, streamflow, and runoff coefficient (ROC) for the 1964-1976 pre-Hugo and 2005-2014 post-Hugo periods are shown in **Figure 4** and **Figure 5**, respectively for this predominantly forested watershed.

The average annual rainfall of 1306 mm for the post-Hugo period was only 1% lower than the pre-Hugo period. However, the average annual outflow of 247 mm in the post-Hugo period was about 20% lower than that for the pre-Hugo period, likely due to increased ET as a result of increased temperature in the region [3] and regenerated pine stands. This flow pattern was evident also from the flow duration analysis presented above where the pre-Hugo frequency of flows < 2.5 $mm \cdot d^{-1}$ occurring >91% of time was higher for the post-Hugo period. The post-Hugo mean annual runoff coefficient of 0.18 was significantly ($\alpha = 0.05$) lower than that (0.23) obtained for pre-Hugo period, potentially due to relatively drier years with higher potential evapotranspiration (PET) [50] resulting in much reduced flows in 2011 and 2012. The seasonal streamflow dynamics of the watershed is much dependent upon the antecedent conditions defined by the water table position or initial flow rate with generally lower flows during the summer growing season with high ET demands compared to the wet winter periods with high water tables as shown in related studies [12] [18] [51].

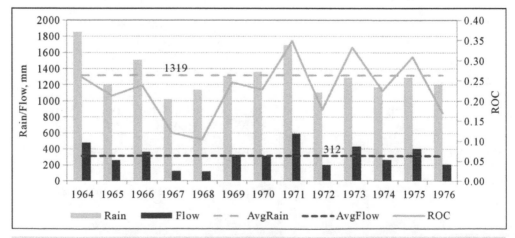

Figure 4. Annual precipitation, stream flow and runoff coefficient (ROC) for 13 years (1964-1976) pre-Hugo period.

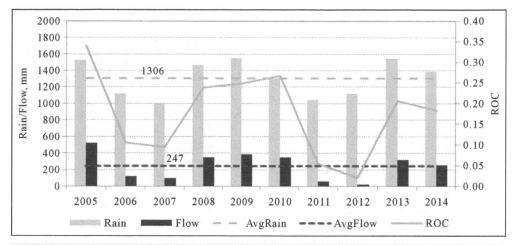

Figure 5. Annual precipitation, stream flow and runoff coefficient (ROC) for 10 years (2005-2014) post-Hugo period.

Using the 13 years of historic data the average annual water yield was 312 mm, which is equivalent to 16.35 × 10^6 m³. On an average annual basis, evapotranspiration (ET), calculated as the difference in precipitation and stream outflow, was estimated to be 1008 mm (after adjusting flow for drainage area in [27]), which is 11% less than the estimated PET of 1135 mm using Hargreaves-Samani-based PET corrected for monthly adjustments [3]. However, the average annual water yield for the last 10-years (2005-2014) was calculated to be only 12.94 35 × 10^6 m³, a reduction in 20% compared to the pre-Hugo estimate. The 10-year data indicate that the average annual water yield of this watershed may possibly be in decline. With respect to extractions of surface water for drinking water supply or irrigation, there aren't any records and it is unlikely that surface water is being used for these purposes. The extent and continuing function of hydrologic modifications have not been addressed and may be contributing to increased peak flows and reduced baseflows.

Hydrology of this forested watershed is complex due to the low-gradient stream network and wide riparian floodplain or linear depressions in the landscape that may flood, but not necessarily from overbank flow. Accurate measurement of stream discharge for extremely large events is complicated by shallow flooding across large areas with variable roughness and braiding of channels. Furthermore, the existence of weakly to strongly connecting wetland swamps, swamp streams and riparian floodplains along streams makes the interaction of surface and subsurface water even more complicated because of their dominance within the landscape as well as spatial and temporal variability within the watershed [52] and considering groundwater-surface water interactions in wetlands for integrated water resources management [53]. Difficulty in accurate estimation of watershed boundary in this low-gradient system also poses challenges in estimation of depth-based watershed stream outflows, nutrient exports, and in design of engineering structures.

3.3. Hydrogeology and Groundwater Resource Data

The USGS has compiled a national atlas of groundwater in the United States and published detailed regional studies of all major aquifers, including the Southeast US [35]. Also, the state agency South Carolina Department of Natural Resources, within their Land, Water and Conservation Division, monitors the deeper confined aquifers in the Coastal Plain region [54] [55]. As stated above in the Physical Data Section, there are now updated geological information for sediments comprising the shallow sediments nominally referred to as the surficial aquifer in the coastal plain in South Carolina [5]. Some local shallow wells were used to perform slug and bail tests using the volume displacement method [56] [57]; the hydraulic conductivity of the surficial aquifer in the northwestern area of Turkey Creek watershed is about 1×10^{-6} m·s^{-1} for the silt loams of the Craven series and 4×10^{-6} m·s^{-1} for the sandy clay loams of the Wahee series.

Except for shallow water-supply wells in the surficial unconfined sediments, which typically produce less than 100 liters per minute, the main groundwater resource for individuals and perhaps smaller communities in Berkeley County is the Santee Limestone, an Eocene-aged, semi-consolidated silty limestone and part of the same lower Floridan aquifer system that is an important groundwater resource in coastal Georgia and in Florida

[58]. However, urbanizing areas have transitioned to surface water sources to avoid further drawdown of groundwater levels and the potential for salt water intrusion into fresh groundwaters. The Floridan/Santee aquifer is partially-confined beneath the Turkey Creek watershed, where the top of the Santee Limestone is approximately 20 m below ground surface (bgs) in the western side of the watershed and about 13 m bgs in the eastern area, and further east at the Santee River, the eastern boundary of FMNF, it is close to or at ground surface. Although not directly measured at the Turkey Creek area, the depth ranges in thickness from 60 - 120 m [59]. In the western area the Santee Limestone is overlain by about 10 m of dense, semi-consolidated calcilutite known as the Parkers Ferry Formation of late Eocene age [60]. Initial groundwater data at the site suggest this unit acts as a confining layer to groundwater flow between the shallow surficial sediments and the deeper Santee Limestone. However, it is not present in the eastern area of the Turkey Creek watershed, where fine-grained sands of the Pleistocene-aged Ten Mile Hill (Talbot) beds overlay the Santee Limestone [60]. An ongoing study analyzing hydrographs of hydraulic head in water-table wells and piezometers, using data collected continuously since February 2006 has focused on groundwater recharge in this watershed and baseflow in Turkey Creek to understand the role of shallow groundwater on stream flow behavior [11].

3.4. Water Quality Data

No water quality data were available on the study watershed for the pre-Hugo 1964-1976 period. However, in 2001 water samples were collected from a section of the stream about one km upstream of the current gauging station on Turkey Creek as one of the three locations in Francis Marion National Forest to examine several water quality parameters, including fecal coliform, dissolved oxygen (DO), conductivity (total dissolved solids), mercury and salinity. Sampling was conducted to ensure that forest streams, lakes, wetlands, and riparian areas are healthy, functioning ecosystems that produce sustained flows of high quality water and to provide a reference for future planning purposes [61]. The results from September 2001 to January 2002 indicated that the Turkey Creek stream water was impaired for fecal coliform bacteria (100 colonies per 100 ml).

Electrical conductivity remained relatively stable (<200 $\mu S \cdot cm^{-1}$) through the measurement period (June-August, 2001). Turkey Creek was repeatedly found to have dissolved oxygen (DO) values of less than the threshold value of 5 $mg \cdot l^{-1}$ during the summer of 2001. In terms of salinity, Turkey Creek near the old gauge (about 200 m downstream of the current gauge) maintained a salinity level of 0.1 ppt throughout the sampling period of June-August, 2001, indicating that there is no tidal influence. Because of the heightened concern over the amount of mercury (Hg) and methylmercury in coastal waters, Turkey Creek was also tested for total Hg and methyl-Hg concentrations [61]. The concentrations of Total Hg and methyl Hg in waters of Turkey Creek were 3.2 and 0.34 $ng \cdot l^{-1}$. Total Hg concentrations in water and mud provide some measure of the potential for contamination. The authors reported that because of the proximity of the National Forest that contains Turkey Creek to other known areas of mercury contamination and the abundance of wetlands within the National Forest, some caution should be taken when regularly consuming fish on or adjacent to National Forest land, and particularly those species of fish that are carnivorous, such as bowfin, large-mouth bass and black crappie.

The high stream water temperatures recorded in the summer were consistent with the estimated high PET for relatively dry months. As expected, the DO concentrations showed an inverse relationship with the water temperature with high values during the cold winter and lower values during the hot summer months, with an average of 6.1 $mg \cdot L^{-1}$, consistent with previous findings [61]. The pH values ranged from 5.4 to 8.8 with an average of 7.1 for 2006-2009 with a no definitive pattern [15]. The annual mean pH of all watersheds significantly increased with stream conductivity ($p < 0.05$). Most of the nitrogen (N) component was dominated by the organic N (ON). Phosphate concentration showed a systematic decreasing trend from 2006 to 2012 [62]. Nitrogen and phosphate concentrations varied with amount of rain received with years receiving highest rainfall registering lower concentrations, possibly due to dilution effects. Data of nutrient concentration measured on this watershed are shown to be within the ranges for similar land use of the coastal plain, except for NH_4-N, which was slightly higher. A detailed analysis of stream water chemistry is still underway. These data may be useful as a reference for comparison with more developed watersheds in the region.

3.5. Socio-Economic Data

There are no specific socio-economic data for the Turkey Creek watershed. Socioeconomic information was derived from data for Berkeley County where this watershed is located, with primary employment in fisheries, forestry, and agriculture. The secondary employment in the region is in agro-industry and energy. Potential tertiary services may include construction, commerce and civil employment. According to the US Census Bureau [63], 66.5% of the population 16 years and older are in the labor force, of which 3.2% are employed in the Armed Services. The unemployment rate for the county is 11.9%. The employment distribution as a percentage of population is estimated at 31.0% in management, business, science, and arts; 25.5% in sales and office occupations; 17.1% in service occupations; 11.9% in natural resources, construction, and maintenance occupations; and the remainder in production, transportation, and material moving occupations. The main industries in Berkeley County include educational services, health care and social assistance (19.5%); manufacturing (13.5%); retail trade (11.8%); and professional, scientific, management, and administrative services (11.8%). Recent ongoing expansion of industrialized areas of the Greater Charleston into Berkeley County may influence the current employment statistics.

3.6. Demographic Data

The population and its growth rate for Berkeley County where the watershed is located are 177,845 and 9.1% per decade, respectively, as per the 2010 census [63]. Unfortunately, the population data for the watershed is not available through the census, but this is a rural area dominated by 98% national forest ownership. We conducted a survey of the watershed in May 2006, which revealed 12 houses within the watershed. We estimate total number of 33 people based on average family size per household of 2.77 for Berkeley County in 2010, yielding a population density of 0.52 person·km^{-2}, which is considerably below the regional norm [63]. The life expectancy as a whole for South Carolina is 74.9 years and 80.2% of the population with 25 years of age and higher was found to have high school diplomas or hold higher degrees.

3.7. Water Usage Data

The households within the watershed use groundwater wells for drinking water. Water consumption per capita is estimated to be 300 l·day^{-1} per person based on some unpublished data for rural communities in South Carolina [37]. All houses are equipped with septic tank systems for sewage disposal. The main source of past water supply using groundwater wells has moved in the direction of surface water sources with treatment due to lowering water tables and potential for salt water influx into fresh groundwater. Septic systems are often used in rural areas for the treatment of sewage, but the ability of many soils to provide ample filtering and percolation has been an issue. Community and urban water treatment systems are in most instances the new norm.

3.8. Vegetation Data

The vegetation on the watershed is mostly comprised of mixed pine-hardwood forests. The current forest stands in the watershed include loblolly pine (*Pinus taeda* L.), loblolly pine-hardwood, pond pine (*Pinus serotina*), pond pine-hardwood, longleaf pine (*Pinus palustris* P. Mill.), baldcypress (*Taxodium distichum*), yellow pine (*Tamias amoenus*), Southern oak (*Quercus falcata*)-yellow pine, bottomland hardwood-yellow pine, white oak-red oak (*Quercus falcate*)-hickory, sweetgum (*Liquidambar styraciflua* L)-yellow poplar (*Liriodendron tulipifera*), sweet gum-oak-willow, baldcypress-water tupelo (*Nyssa aquatica* L.), sweet bay (*Magnolia virginiana*)-swamp tupelo (*Nyssa sylvatica*)-red maple (*Acer rubrum*), and flatwoods-brush [32].

4. Water Resource Assessment

4.1. Water Demand Assessment

Since the population on the watershed is so small and depends primarily upon the groundwater wells, no assessment is available and substantial uncertainty exists about past hydrological modifications and their ongoing effects. Accordingly, the only source of withdrawal from the watershed is by forest/vegetation ET, which is estimated at 1030 mm·yr^{-1}.

4.2. Environmental Impact Assessment and Strategic Impact Assessment

No environmental or strategic impact assessments have been conducted specifically for the watershed to date. However, environmental assessments for forest health and stream water quality have been conducted by the

Forest Service for the whole Francis Marion National Forest, which includes the Turkey Creek watershed [61]. As stated in that report, the quality of stream water of Turkey Creek was reported as being impaired for coliform bacteria. The likely sources include wildlife, cattle and horses on the small farms, and septic systems. The invasive wild hogs are a growing coastal menace relative to their riparian and water quality impacts. However, there are no specific study data to document the extent or probable source areas. Another emerging issue is the methyl-Hg, and monitoring has recently been initiated. Currently the draft version of the environmental impact study (EIS) associated with the ongoing National Forest plan revision [64] suggests that Turkey Creek will be impacted more than other vicinity subwatersheds relative to timber thinning and harvesting that favors restoration of longleaf pine and prescribed fire where suitable within the urban interface. Turkey Creek may also be selected as a priority watershed where efforts to improve watershed condition and function will be increased.

4.3. Social Impact Assessment

No specific social impact assessment of management activities on the Turkey Creek watershed has been conducted to date. However, the public review process of forest management practices by the National Forest lands does include social and community considerations. There are also programs to increase the public awareness of prescribed fire and smoke management, restoration of native plant communities, and recreational uses of the forest. The National Forest has recently been recognized for the values it provides by the community leaders, reflecting the importance of forest lands for ecological, economic and cultural values.

4.4. Risk or Vulnerability Assessment

The South Carolina Department of Health and Environmental Control (SCDHEC), under regulation by the 1973 Federal Clean Water Act, posts a listing (called the 303 (d) list) of impaired water bodies that do not meet federal standards, lists the water bodies already impaired or at risk or vulnerable, and the activity restrictions. As mentioned earlier, Turkey Creek was listed under 2006 303 (d) list for exceeded limits of fecal coliform. However, Turkey Creek is not on its current 2014 listing, so the impairment may have been temporary, based on circumstances that have been addressed. Water quality information mentioned above was reconnaissance level data [61]. Research studies suggest that there are very few water quality related issues, but periods of drought cause flow to stop with associated impacts to aquatic and riparian life. The current revitalization effort with multiple monitoring and modeling studies will provide a basis in the near future for a more detailed risk assessment of water quantity and quality. Estimated in the draft forest plan environmental analysis suggest existing activity within the watershed increase water yield by 8% and proposed activity within the next decade may increase water yield between 5% and 11%, based primarily under low density forest management and prescribed fire [64]. These estimates will be compared with results from ongoing SWAT modeling assessment for various climate change scenarios at the study site.

5. Integrated Water Resource Management (IWRM)

5.1. Issue Identification

Many of the ecohydrologic and integrated water management issues on this freshwater system proposed/identified at earlier phase of the project have been addressed by collaborative approaches during last 10 years as reported recently [7]. For example, reports have been generated concerning issues on design peak and low flow discharges for various design storms and flow duration pattern including their scaling issues generally used in pre-development storm water management scenarios, design of cross-drainage structures, and allocation of environmental flows [44]; water budget components generally used in restoration and comparison with developed watersheds [3] [14] [15] [17] [30] [65]; storm event characterization in terms of their peak flow rates, time to peak, runoff volumes and duration, and runoff ratios also used in pre-development scenarios and design of the best management practices [12] [18] [51] [66]; and storm runoff generation process and contributions of the subsurface base flow and surface runoff in total stream yields needed for water management and estimating pollutant discharges [11] [12] [15] [17] [18] [51] [67]. A study on recharge rates into the underlying groundwater aquifer and its quality in this region was conducted [11], related publications on the interaction of precipitation, surface and groundwater flow and their relative contributions to stream flow dynamics in the riparian floodplain of this lowland forested watershed are also available [67]-[69].

The issue with resolution of available topographic maps and associated digital elevation models (DEMs) in delineating watershed boundary for the drainage area and estimating effective surface depressional storage representing heterogeneity in surface micro-topography on hydrologic functions were also studied [34] [70]-[72]. Issues related with effects of climate variability and potential climate change including extreme natural disturbance like Hurricane Hugo in 1989 on vegetation dynamics, long-term streamflow, ET, and nutrient exports of these watersheds have also been studied and reported [3] [14] [15] [73]-[76]. Many of those above studies provided data and information in terms of water resources management on the Turkey Creek watershed primarily based on characterizing the watershed as a baseline "reference" for inter-site comparisons, evaluating the impacts of other developing watersheds in the region, and designing the best management practices. However, there is some uncertainty with applying Turkey Creek as a reference stream that must be considered. Little is known with respect to the differences that exist relative to marine terraces and past hydrologic modifications. For example, substantial variability of drainage size exists in the NC coastal plain before developing intermittent or perennial streams [77]. Without substantial more information and field verification, our assumption must be that conditions are variable and continuing efforts will be needed to refine and clarify coastal plain hydrology. Streams within each geologic terrace may be affected differently, such as how deep erosion has entrenched into the marine terrace surface, or how deep restrictive layers are such as identified by shallow water tables from estuarine silt and clay deposits. There are too few long-term coastal stream gauging stations and ongoing research to know how far we have come, let alone how far we must go to understand coastal systems. The thought is that we do now know substantially more than we did thirty years, even 10 years ago. However, due to increasing urbanization and population growth near the coastal waters together with changing climate and sea level rise, newer emerging issues on storm water management, increased demand of water supply and quality, management for ecological restoration, and other associated ecosystem services will continue to be complex and a challenge to address. Therefore, additional science based collaborative research on integrated water management should continue to address such challenges including some of other key science questions related to this low-gradient coastal forest ecosystem listed below:

What are the uncertainties associated with current knowledge of surface and groundwater interactions and their partitioning that affect the key ecosystem services like water supply, water quality, and biogeochemical processes and transport in this coastal landscape? How will those uncertainties be addressed?

What is/are the cause(s) of the high methylization efficiency of Hg in these streams, and how do forest management practices affect Hg transport and methylization? How will the processes affect at this site compared to other studies?

How will a changing climate affect the hydrology (particularly streamflow and ET) and water quality of this forest ecosystem in next decades and what will be their implications to the current forest ecosystem? What are the spatial and temporal dynamics of ET as affected by various stand types on the watershed?

In what ways are the presence of the existing culverts and bridges and legacy water management structures influencing the runoff and its quality and pathways, peak discharges, timing and baseflow, especially in the face of changing climate where high intensity storm events are in forecast for the region? [78]

To what extent may the surface and groundwater hydrology and thus the riparian vegetation and aquatic habitat be influenced by potential sea level rise as a result of climate change?

What are the mechanisms and rates of feedback between topography, hydrology, ecology (vegetation), and geochemistry and how do these vary as a function of local climate? Are there threshold phenomena in these feedback mechanisms and can the mechanisms be quantified?

How does the interaction of tidal waves downstream of the Turkey Creek watershed, which discharges freshwater outflow to Huger Creek downstream, affect the eco-hydrological and biogeochemical processes including net primary productivity, carbon balance and greenhouse gas emissions in the tidal streams in the face of changing climate and sea level rise? [79]

How do we develop research techniques to minimize landscape scale eco-hydrological effects of forest management treatments like prescribed fire or biomass treatment, fuel hazard reduction generally applied for improving the forest health in this landscape?

5.2. Defining Management Options-Scenarios

The National Forest is revising the existing management plan for the forest. Within that context, forest roads, timber harvest, prescribed fire, sustaining habitats, hiking trails, streams, riparian buffers and wetlands are also included. The development of the forest plan considers management options, conservation needs and public use, with each option subject to public review. The plans are updated on approximately a 10-year basis, unless specific needs arise, situations change, or regulations are modified. Currently a new 10-year plan (2016-2026 or beyond) forest management draft is underway with inputs from the stakeholders. There are also supplemental assessments, which are specific to a particular use or management objective. For example, a special roads analysis [80] was prepared as a supplement to the Francis Marion National Forest Land and Resource Management Plan of 1996, to address the use and needs of forest roads. In the recently prepared draft, two key issues that are being addressed are: 1) to what extent and where should native ecological systems be restored? and 2) what is the best approach for dealing with the rapid change in land use and population growth? [64]. In that draft, Turkey Creek was identified as one of the priority watersheds for improving hydrologic functions based on historic past modifications as well as the rehabilitation of existing cross-drainage structures that affect wetland and riparian structure, biota, processes, and functions and restoring longleaf pine ecosystem together with its red-cockaded woodpeckers habitat as an at-risk species. In making this consideration, it is likely to also be a center of forest management activity with extensive thinning and prescribed burning for forest health, support of ecological systems and to address forest health and fuel loading.

All management options include an array of forest management practices with varying intensities of stand tending including regeneration, thinning and whole-tree biomass removal or mulching (both understory and overstory). Removal of thinned biomass as a management treatment is undertaken primarily to reduce fire hazards and provide a green solution to augment burning fossil fuels. To a great extent—after Hurricane Hugo— the forest has matured with commercial value too expensive to purchase and burn. However, understory biomass removal using mastication or prescribed burning may still be conducted for restoration of longleaf pine and in areas where burning fuels is unacceptable. Without fire disturbance, loblolly pine is the more competitive, faster growing species. Both species are facultative species that can be found in wetlands including wet pine sites. Longleaf pine favors prescribed burning and sustains ecosystems with higher ecological diversity and longevity, more resilient to disturbances such as those associated with climate change. Furthermore, restoration of longleaf pine ecosystems helps to restore the endangered, red-cockaded woodpecker (RCW) habitat. Following the stand tending operation, prescribed fire is conducted generally every two to three years to maintain a relatively low stocking level in the understory. Also, as the National Forest manages this public land for recreational purposes such as hiking, hunting, horse riding, fishing, bird-watching, canoeing, driving all-terrain vehicles (ATVs), management decisions are often taken in terms of keeping the trails and land safe and pleasant. Due to the sensitivity of some species, activities have to be constrained to limit their impact to only designated areas.

5.3. Establishment of Decision Criteria: Constraints and Factors

The USDA Forest Service is the decision maker for actions on its national forests. Those decisions are subject to local, state and federal laws and regulations, and there is coordination with state and federal agencies. There are numerous laws, executive orders and regulations that directly or indirectly address some aspect of water management on the National Forests [81]. Policy instruments for water quantity, quality, ecosystem habitat, and spatial planning are available for managing the water system and a variety of other resources. Similarly, the general public and stakeholders (landowners on the watershed) have collaborative roles within the environmental assessment, analysis and decision making process. The National Forest maintains technical staffing support to help prepare draft documents and line officers with designated authority within the agency review documents and make decisions.

The decision criteria for water resources management as such have not yet been formulated for this watershed to our knowledge. However, regardless of the decision, the alternatives suggest there will be some emphasis on managing and restoring forest species (particularly longleaf pine), wildlife habitat and increased considerations given to hydrologic modifications that may benefit water resources. For example, areas of thinning, longleaf conversion and prescribed burning in the watershed may reduce transpiration and increase soil saturation possibly resulting in increased water quantity (stream flow) and stream water nutrient concentrations on a short-term basis [14] [82]. The increasing emphasis on frequent prescribed burning to maintain woodland and savanna densities will not only help sensitive habitats, but also help to maintain water yield increases. One of the constraints for such large-scale activities is to protect water quality and improve aquatic habitats with the riparian management zone buffers while also considering activities to protect wetland and riparian functions outside of this zone. Prescribed fire may still occur within the riparian areas, but greater attention is made to allow the prescribed fire to back into riparian areas rather than direct lighting. A recent case is with the adjacent first-order watershed (WS 77) (**Figure 1**) where 84% of the land had understory burning on May 10, 2003 leaving only the riparian buffers along the stream. Hardwood buffers along streams supply better food for aquatic organisms, and they also reduce fire frequency and intensity as compared to pine trees. Burning may also be limited near the urban interface near Highway 41, the downstream boundary of the gauged watershed. The prescribed burning is becoming a larger issue as the urbanization of Charleston has been expanding and the watershed as well as most of the National Forest is located near or within the wildland-urban interface (WUI).

Any new activities and studies related with the revitalization of the monitoring of Turkey Creek watershed may provide decision support information. Accordingly, the Turkey Creek watershed research and study projects are being planned and implemented as a multi-collaborative approach considering both the needs for additional information as well as availability of resources.

5.4. Data Acquisition

The gauging station was reactivated in 2005, with the measurement of stream flow and rainfall as described above (see Sections 1, 2, and 3.2; http://waterdata.usgs.gov/sc/nwis/uv?site_no=02172035). Stream flow rates continue to be obtained by using stage heights using a Sutron datalogger connected to the pressure transducer at the bottom of the stream with a stage-discharge relationship developed by the USGS using frequent *in situ* manual velocity measurements with a Marsh-McBirney flowmeter. Rainfall is also measured with an automatic tipping bucket at the stream gauging station. A Campbell Scientific, Inc. weather station with a CR-10X datalogger was installed upstream in the watershed in October 2005 to monitor precipitation, air temperature, relative humidity, wind speed, and solar radiation). Later in August 2006, with support from the FMNF, four additional shallow wells (to depth of 3 m) were also installed in four different soil types to monitor the water table on the south side of Turkey Creek watershed. In 2010, additional shallow groundwater well was installed on Wahee soil on the north side of Turkey Creek. Since late 2010 two smaller tributaries have been monitored for streamflow.

Stream water quality monitoring on this watershed was initiated with a bi-weekly grab sampling at the new gauging station in October, 2005 that was augmented later in June 2006 with the addition of a flow proportional sampling of storm events using an ISCO-4200 automatic sampler [83]. Both the manual grab and flow proportional samples for nutrients (NO_3-N, NH_4-N, Total N, and PO_4) and some other cations (Ca^+, K^+, Na^+), and Cl^- from the ISCO auto samplers were analyzed in the Center's Soil and Water Chemistry laboratory in Charleston, SC until mid-2007 after which analysis is being conducted by the USDA Forest Service Coweeta Hydrologic Laboratory in Otto, North Carolina. Water quality for physical parameters (dissolved oxygen (DO), pH, temperature, turbidity, salinity, and conductivity) are measured *in-situ* every 3 - 4 weeks using a HYDROLAB's Eureca multi-sensor probe began in June 2006 also on a bi-weekly basis. Detailed procedures of the hydro-meteorologic and water quality monitoring have been described elsewhere [15] [18] [30] [32] [68] [83].

With the cooperation of College of Charleston, in December 2005 three sets of piezometers (a cluster of three groundwater wells) were installed along a cross-forest transect from northwest to southeast. One set was installed within the Turkey Creek watershed (adjacent to the weather station, and screened at depths below ground surface of 4.5 m, 6 m, and 16 m, respectively) and the other two piezometer sets were located outside the watershed. One piezometerset was installed in the Santee Experimental Forest watershed approximately 2 km NNW of the USGS Turkey Creek stream gage; the third piezometer set was installed about 15 km to the SE of the Turkey Creek stream gage [11].

Geographical Information Systems (GIS) data on topography from digital elevation models (DEMs), hydrography (drainage network), soils, vegetation, land use, and land management for the watershed are compiled primarily from the web sites of the SC DNR (www.dnr.sc.gov/GIS/gisdownload.html), and the Francis Marion National Forest (www.fs.fed.us/r8/fms/forest/Services/GIS/export.html). Similarly, aerial photographs for the study area were acquired from the National Forest for increments of five years starting in 1963, 1968, 1973, 1978, 1983, 1989 (before and after Hurricane Hugo), 1999 and thereafter are available from the National Forest). Preliminary work has been undertaken to refine the subwatershed boundaries and stream networks using LiDAR and high density aerial photography within the Francis Marion National Forest [33]. Ongoing work within the

FMNF includes conducting and inventory of stream crossings and stream channels to provide improve aquatic information. A field survey was completed on the major tributaries and main channel including the road culverts and bridges using a surveyor's Trimble Global Positioning System (GPS) unit [32]. In mid-2011, LiDAR data for the study watershed were obtained from the Berkeley County dataset (J. Scurry, personal communication, 2011). The 1.5 m DEM created from the LiDAR data along with extensive field verifications for cross-drainage culverts along the roads bordering the watershed was used to obtain the new watershed drainage area of 5240 ha [34]. All other types of data have been described above in Sections 3.2 to 3.8. Data on hydro-meteorology, water quality, and land resources are now available through GIS-based data sharing web-portal at the site: http://www.srs.fs.usda.gov/charleston/santee/data.html. Continued monitoring on this watershed in coming years will provide important information on long-term water budget, water yield, flooding pattern, and surface-subsurface water interactions needed for water management and appropriate ecological decisions. Furthermore, these data also help to better understand hydrologic effects of land use change and forest regeneration, after the impact of 1989 Hurricane Hugo [36]. Studies are already underway to examine seasonal storm event characteristics and dynamics, water budget, and ET of the forest ecosystem using historic data. Results from these studies will provide baseline information for an adjacent watershed (Quinby Creek) as well as be useful for comparison with other watersheds in the physiographic area. Current measurements on rainfall, weather, stream outflows, shallow and deep groundwater levels are providing data for further studies of the water balance, surface-subsurface water interactions, and the interactions of rainfall and soil types on shallow water table dynamics and stream outflow. Eventual correlations of moisture and soil data should help improve implementation guidance for ground disturbing activities. Understanding groundwater-surface water interactions will help the water resource managers deal with issues such as flood mitigation, groundwater exploitation, and biodiversity conservation, in a more integrated and sustainable manner [53]. A short term field sampling study to examine the impacts of bridge surface runoff at the Turkey Creek gauging station on the stream water quality downstream is underway in collaboration of the USGS with South Carolina Department of Transportation.

5.5. Decision Support Process

A physically-based, spatially distributed, watershed eco-hydrologic model, Soil and Water Assessment Tool (SWAT) [45] [84], was parameterized as a decision support tool for the Turkey Creek watershed [30] [32]. This model is capable of integrating various hydrological, ecological and biogeochemical processes occurring on a field-scale that can be extended into a large watershed/basin scale [85]-[87]. The simulation model was calibrated and validated using the watershed spatial data as well as temporal and spatial data on precipitation, weather, stream flow rates collected through 2010, and groundwater table. The validated model can then be used for its further application to evaluate management decisions related with anthropogenic and natural disturbances. The SWAT model has been used as a decision supporting tool to evaluate impacts of land use and climate change and management and basin wide water management [88] [89].

5.6. Open Access and Transparency of Output

The parameterized model, data for calibration, and simulation results will be made accessible to others including the public after a proper quality assurance review. This reporting will be implemented through web posting http://www.srs.fs.usda.gov/charleston/, reporting in cooperators' meetings, meetings and conferences of professional societies, progress reports to the cooperators, in addition to publication in conference proceedings and journal papers, and information bulletins.

5.7. Integrated Modeling Approach

It is very important that an integrated model be used to incorporate physically-based hydrologic, biogeochemical and ecological processes that are applicable to poorly drained, high water table soils followed by the flow routing schemes that consider wetland depressions, braided streams, backwater and tidal effects. Hydrologic processes on relatively low-gradient poorly drained coastal plain sites, like Turkey Creek watershed, are usually dominated by wetlands with variable shallow water table positions. Most of the outflows (surface runoff and subsurface drainage) from these watersheds drain from saturated areas where the water is at the surface or a shallow water table is present meaning that the otal outflow depends in part on the frequency and duration of flooding and also on the dynamics of the water table (hydroperiod), which are driven by rainfall substrate permeability and evapotranspiration (ET) [90]. We have chosen the SWAT model [85] as an integrated comprehensive model that has been extensively tested and applied for modeling runoff and evapotranspiration (water budget), erosion and sediment transport, groundwater, river hydrology, flooding (flood management), water chain, water quality (nutrients), evapotranspiration, ecological-environmental (harvesting, burning etc.), and economical parameters. For example, [84] recently demonstrated the applicability of the SWAT model in predicting daily stream flows from coastal Louisiana watersheds, which have primarily forested and agricultural land uses. [91] also used the SWAT to model point and diffuse source pollution of nitrate in a rural lowland catchment.

The model is currently being further refined using the new LiDAR-based, high resolution DEMs for subwatershed boundary and stream tributary delineation and other related model input parameters using the ArcSWAT9.3 platform. Available concept of HRUs (Hydrologic Response Unit) in the SWAT model is also used for the areas with specific management such as the open areas and private lands, wetlands for accurately assigning the parameters and simulating the processes. Hydraulic parameters needed for flow routing process were obtained from both the literature and also the stream and drainage network survey at the site. However, some uncertainty may exist for braided streams potentially with higher roughness. Soil hydraulic properties for major soil types are being used either from the measurements or from the published data for the region. Rainfall data is being used from three automatic gauges within and just outside the watershed. Data from the *in-situ* weather station as well as nearby station above forest canopy (WS 80) in a tower are being used for forest potential ET (PET) estimates. All other vegetation parameters were obtained either from the literature or some field measurements. The reliability of the SWAT model in predicting the stream flow processes on poorly drained soils published earlier by [30] on this watershed is being tested further. Results so far have been favorable. Furthermore, development of an analytical model to estimate daily water table dynamics as a function of daily rainfall, PET, and soil and vegetation parameters for major soil types on the watershed is underway.

6. Developing Water Management Indicators

Hydrologic and water management indicators are standard metrics obtained by measurements of streamflow and water quality sampling of chemical (NO_3-N, NH_4-N, Total P, Total N, DOC, Cl, Ca, K, Mg, and Na) and physical (pH, DO, temperature, turbidity, salinity, and specific conductivity) parameters. Stream flow rates can be analyzed to obtain the design maximum and minimum flow, flow frequency duration, mean runoff coefficient, water yield and storage including their variability. Alternatively, the proposed "Indicators of Hydrologic Alteration (IHA)" method is based upon an analysis of hydrologic data available either from existing measurement points within an ecosystem (such as at stream gauges or wells) or model-generated data [49]. In this method 32 parameters, organized into five groups, to statistically characterize hydrologic variation within each year provide information on ecologically significant features of surface and ground water regimes influencing aquatic, wetland, and riparian ecosystems. Results of a recently conducted survey of fish species counts can be found elsewhere [92], which may also serve as a stream biological indicator.

7. Ecosystem Assessment

The condition of the forest ecosystem within the watershed is assessed on a periodic basis through the forest planning and revision process described above and watershed condition assessment of national forest subwatersheds by the forest team of resource specialists [29]. In addition, aerial surveys are occasionally conducted and photos taken, when needed, to assess a) insect and disease outbreaks, b) changes in forest condition or cover, such as red-cockaded woodpecker habitat, and c) wildfire patrols. The physical, biological and cultural assessment approaches have been developed for each resource area by the National Forest to be consistent with law, regulation and quality resource management. All of the above data from this relatively undisturbed forested watershed can possibly be used as a "reference" for eco-hydrological and water quality assessments of developing watersheds in the coastal region. Additional knowledge produced from future interdisciplinary collaborations, coupled with the application of modern modeling techniques, will determine the accuracy and precision of future ecohydrological assessments [93].

8. Summary

This case study provides a comprehensive water resources assessment in terms of surface water and groundwater as related to the soils, geology, and vegetation for a 5240 ha forested watershed, one that is typical to the Atlantic Coastal Plain. The study also provides information on existing water resources knowledge including modeling integrated water resources management based on last 10 years of collaborative studies, administrative policy and decision process, and socio-economics as related to water resources management on this type of watershed assessment. Various types of data needs and acquisition including applications of socioeconomics were highlighted to respond to more complex, broader scale problems with cumulative ecohydrologic effects using an integrated watershed management approach on a larger watershed. The long-term eco-hydrologic data available for the study watershed after its revitalization in 2005 provided a sound scientific basis for multidisciplinary collaborators to advance the understanding of processes, characteristics, and their interactions on the lower coastal plain landscape, including its resiliency after an extreme hurricane event (Hugo) that occurred in 1989. Although the data and information generated from this study may be useful as a "reference" for comparison of coastal water resources and management issues, it is becoming increasingly clear that one example is helpful, but multi-site studies may be warranted to understand these complex systems in the face of climate change and sea level rise. The need for further research in several areas was highlighted, including 1) accurate partitioning of surface and subsurface flow, 2) ET dynamics, 3) monitoring for mercury and fecal coliform, 4) developing techniques to improve the forest health such as by prescribed fire or biomass treatment for fuel hazard reduction, 5) assessing ecosystem impacts of climate change and others to address the contemporary issues on this and other similar coastal watersheds.

Acknowledgements

Initial support for reestablishing the gauging station and the associated data collection, management, and analyses was provided by the USDA Forest Service and the National Council for Air and Stream Improvement, Inc. (NCASI). The reactivation of the watershed monitoring has been conducted in cooperation with the US Geological Survey, the College of Charleston, and the South Carolina Department of Transportation. This work was supported by the USDA Forest Service and partially by the NATO Committee on Challenges for Modern Society (CCMS) Integrated Water Management Program (IWM), University of Antwerp, Belgium. Authors would like to acknowledge Francis Marion National Forest, US Geological Survey, College of Charleston, South Carolina Department of Transportation, Agricultural University of Krakow and Warsaw University of Life Sciences in Poland, Clemson University, Florida A&M University, for their various levels of support in this project. Thanks are also due to Lisa Wilson, Bray Beltran, Ileana La Torre Torres, Elizabeth Haley, former graduate students, respectively, at the College of Charleston, and Andy Harrison, Julie Arnold, and Matt Kraswoski (formerly), USDA Forest Service, Center for Forested Wetlands Research, for installing monitoring stations and data collection. The authors also express their thanks to USDA Forest Service Southern Research Station Civil Rights Committee for supporting travel grants to Joseph Amoah former PhD student at Florida A&M University and Jose Martin former student at Gainesville State College, GA. Thanks are also due to Jose Martin for completing the LiDAR data processing for watershed delineation and his academic advisor Dr. Sudhanshu Panda for needed guidance. We also highly appreciate Dr. Paul Conrads at US Geological Survey and Dr. Daniel Hitchcock at Clemson University for their constructive comments in their review of this paper.

References

[1]   Reice, S. (2005) Ecosystems, Disturbance and the Impact of Sprawl. In: Johnson, E.A. and Klemens, M.W., Eds., *Nature in Fragments: The Legacy of Sprawl*, Columbia University Press, New York, 90-108.

[2]   Smith, H.R. (2012) Rich Swamps and Rice Grounds: The Specialization of Inland Rice Culture in the South Carolina Low Country, 1670-1861. PhD Dissertation, The University of Georgia, Athens.

[3]   Dai, Z., Trettin, C. and Amatya, D.M. (2013) Effects of Climate Variability on Forest Hydrology and Carbon Sequestration on the Santee Experimental Forest in Coastal South Carolina. SRS-172, USDA Forest Service, Asheville, 32.

[4]   Golden, H.E., Lane, C.R., Amatya, D., Bandilla, K., Raanan-Kiperwas, H., Knightes, C.D. and Ssegane, H. (2014) Hydrologic Connectivity between Geographically Isolated Wetlands and Surface Water Systems: A Review of Select Modeling Methods. *Environmental Modeling & Software*, **53**, 190-206.

http://dx.doi.org/10.1016/j.envsoft.2013.12.004

[5]   Doar, W.R. (2014) The Geologic Implications of the Factors That Affected Relative Sea-Level Positions in South Carolina During the Pleistocene and the Associated Preserved High-Stand Deposits. PhD Dissertation, University of South Carolina, Columbia.

[6]   Messina, M.G. and Conner, W.H. (1997) Southern Forested Wetlands: Ecology and Management. CRC Press, Boca Raton.

[7]   Amatya, D.M., Callahan, T.J. and Trettin, C.C. (2015) Synthesis of 10 Years of Hydrologic Studies on Turkey Creek Watershed. *Proceedings of the 5th Interagency Conference on Research in the Watersheds*, Charleston, 2-5 March 2015.

[8]   Amatya, D.M., Skaggs, R.W. and Gregory, J.D. (1997) Evaluation of a Watershed Scale Forest Hydrologic Model. *Agricultural Water Management*, **32**, 239-258. http://dx.doi.org/10.1016/S0378-3774(96)01274-7

[9]   Aust, W.M., Reisinger, T.W., Burger, J.A. and Stokes, B.J. (1993) Soil Physical and Hydrological Changes Associated with Logging a Wet Pine Flat with Wide-Tired Skidders. *Southern Journal of Applied Forestry*, **17**, 22-25.

[10] Amatya, D.M., Skaggs, R.W. and Gregory, J.D. (1996) Effects of Controlled Drainage on the Hydrology of Drained Pine Plantations in the North Carolina Coastal Plain. *Journal of Hydrology*, **181**, 211-232. http://dx.doi.org/10.1016/0022-1694(95)02905-2

[11] Callahan, T.J., Vulava, V.M., Passarello, M.C. and Garrett, C.G. (2012) Estimating Groundwater Recharge in Lowland Watersheds. *Hydrological Processes*, **26**, 2845-2855. http://dx.doi.org/10.1002/hyp.8356

[12] Epps, T.H., Hitchcock, D.R., Jayakaran, A.D., Loflin, D.R., Williams, T.M. and Amatya, D.M. (2013) Characterization of Storm Flow Dynamics of Headwater Streams in the South Carolina Lower Coastal Plain. *Journal of the American Water Resources Association*, **49**, 76-89. http://dx.doi.org/10.1111/jawr.12000

[13] Slattery, M.C., Gares, P.A. and Phillips, J.D. (2006) Multiple Modes of Storm Runoff Generation in a North Carolina Coastal Plain Watershed. *Hydrological Processes*, **20**, 2953-2969. http://dx.doi.org/10.1002/hyp.6144

[14] Amatya, D.M., Miwa, M., Harrison, C.A., Trettin, C.C. and Sun, G. (2006) Hydrology and Water Quality of Two First Order Forested Watersheds in Coastal South Carolina. *ASABE Annual Conference*, 22.

[15] Amatya, D.M., Callahan, T.J., Trettin, C.C. and Radecki-Pawlik, A. (2009) Hydrologic and Water Quality Monitoring on Turkey Creek Watershed, Francis Marion National Forest, South Carolina. *ASABE Annual Conference*, Reno.

[16] Burke, M.K. and Eisenbies, M.H. (2000) The Coosawhatchie Bottomland Ecosystem Study a Report on the Development of a Reference Wetland. SRS-38, USDA Forest Service, Asheville, 64.

[17] Harder, S.V., Amatya, D.M., Callahan, T.J., Trettin, C.C. and Hakkila, J. (2007) Hydrology and Water Budget for a Forested Atlantic Coastal Plain Watershed, South Carolina. *Journal of the American Water Resources Association*, **43**, 563-575. http://dx.doi.org/10.1111/j.1752-1688.2007.00035.x

[18] La Torre Torres, I.B., Amatya, D.M., Sun, G. and Callahan, T.J. (2011) Seasonal Rainfall-Runoff Relationships in a Lowland Forested Watershed in the Southeastern USA. *Hydrological Processes*, **25**, 2032-2045. http://dx.doi.org/10.1002/hyp.7955

[19] McCarthy, E.J., Skaggs, R.W. and Farnum, P. (1991) Experimental Determination of the Hydrologic Components of a Drained Forest Watershed. *Transactions of the American Society of Agricultural Engineers*, **34**, 2031-2040. http://dx.doi.org/10.13031/2013.31833

[20] Pyzoha, J.E., Callahan, T.J., Sun, G., Trettin, C.C. and Miwa, M. (2008) A Conceptual Hydrologic Model for a Forested Carolina Bay Depressional Wetland on the Coastal Plain of South Carolina, USA. *Hydrological Processes*, **22**, 2689-2698. http://dx.doi.org/10.1002/hyp.6866

[21] Riekerk, H., Jones, S.A., Morris, L.A. and Pratt, D.A. (1979) Hydrology and Water Quality of Three Small Lower Coastal Plain Forested Watersheds. *Proceedings of the Soil Crop Science Society of Florida*, **38**, 105-111.

[22] Skaggs, R.W., Chescheir, G.M., Fernandez, G.P., Amatya, D.M. and Diggs, J. (2011) Effects of Land Use on Soil Properties and Hydrology of Drained Coastal Plain Watersheds. *Transactions of the ASABE*, **54**, 1357-1365. http://dx.doi.org/10.13031/2013.39037

[23] Sun, G., Callahan, T.J., Pyzoha, J.E. and Trettin, C.C. (2006) Modeling the Climatic and Subsurface Stratigraphy Controls on the Hydrology of a Carolina Bay Wetland in South Carolina, USA. *Wetlands*, **26**, 567-580. http://dx.doi.org/10.1672/0277-5212(2006)26[567:MTCASS]2.0.CO;2

[24] Sun, G., Lu, J., Gardner, D., Miwa, M. and Trettin, C.C. (2000) Water Budgets of Two Forested Watersheds in South Carolina. *Spring Specialty Conference of American Water Resources Association*, Middleburg, 199-202.

[25] Young, C.E. and Klawitter, R.A. (1968) Hydrology of Wetland Forest Watersheds. *Proceedings of Hydrology in Water Resources Management*, Clemson, 29-38.

[26] Young, C.E. (1964) Precipitation-Runoff Relations of Small Forested Watersheds in the Coastal Plain. Wetland Hydrology Study Plan W-3, USDA Forest Service, Charleston, 23.

[27] Amatya, D.M. and Trettin, C.C. (2007) Annual Evapotranspiration of a Forested Wetland Watershed, South Carolina. *ASABE Annual Conference,* 4.

[28] Eidson, J.P., Lacy, C.M., Nance, L., Hansen, W.F., Lowery, M.A. and Hurley, N.M. (2005) Development of a 10- and 12-Digit Hydrologic Unit Code Numbering System. USDA Natural Resources Conservation Service, 38.

[29] USFS (2011) Watershed Condition Framework Technical Guide. FS-978, USDA Forest Service, 49.

[30] Amatya, D.M. and Jha, M.K. (2011) Evaluating Swat Model for a Low-Gradient Forested Watershed in Coastal South Carolina. *Transactions of the ASABE,* **54**, 2151-2162. http://dx.doi.org/10.13031/2013.40671

[31] BCD-COG (2000) Water Quality Management Plan, Vol. II. A Report. Berkeley Charleston Dorchester Council of Governments, North Charleston, 11.

[32] Haley, E.B. (2007) Field Measurements and Hydrologic Modeling of the Turkey Creek Watershed, South Carolina. MS Thesis, College of Charleston, Charleston.

[33] Maceyka, A. and Hansen, W.F. (2015) Enhancing Hydrologic Mapping Using Lidar and High Resolution Aerial Photos on the Francis Marion National Forest in Coastal South Carolina. *Proceedings of the 5th Interagency Conference on Research on Watersheds,* Charleston.

[34] Amatya, D.M., Trettin, C., Panda, S. and Ssegane, H. (2013) Application of Lidar Data for Hydrologic Assessments of Low-Gradient Coastal Watershed Drainage Characteristics. *Journal of Geographical Information System,* **5**, 175-191. http://dx.doi.org/10.4236/jgis.2013.52017

[35] Campbell, B.G. and Coes, A.L. (2010) Groundwater Availability in the Atlantic Coastal Plain of North and South Carolina. Professional Paper 1773, US Geological Survey, Reston, 241.

[36] Hook, D.D., Buford, M.A. and Williams, T.M. (1991) Impact of Hurricane Hugo on the South Carolina Coastal Plain Forest. *Journal of Coastal Research,* **8**, 291-300.

[37] SCDNR (2009) South Carolina Water Resources Assessment. South Carolina Department of Natural Resources, Columbia.

[38] Barker, R.A. and Pernik, M. (1994) Regional Hydrology and Simulation of Deep Ground-Water Flow in the Southeastern Coastal Plain Aquifer System in Mississippi, Alabama, Georgia, and South Carolina. Professional Paper 1410C, US Geological Survey.

[39] Colquhoun, D.J. (1974) Cyclic Surficial Stratigraphic Units of the Middle and Lower Coastal Plain, Central South Carolina. In: *Post-Miocene Stratigraphy, Central and Southern Atlantic Coastal Plain,* Utah State University Press, Logan.

[40] NRCS (1980) Soil Survey Report of Berkeley County, South Carolina. US Department of Agriculture, Natural Resources Conservation Service. USDA Natural Resources Conservation Service, 94.

[41] Law, D. (2006) USDA Forest Service. Columbia.

[42] Hargreaves, G.H. and Samani, Z.A. (1985) Reference Crop Evapotranspiration from Temperature. *Applied Engineering in Agriculture,* **1**, 96-99. http://dx.doi.org/10.13031/2013.26773

[43] Monteith, J.L. (1965) Evaporation and Environment. *Symposia of the Society for Experimental Biology,* **19**, 205-234.

[44] Amatya, D.M. and Radecki-Pawlik, A. (2007) Flow Dynamics of Three Experimental Forested Watersheds in Coastal South Carolina, U.S.A. *Acta Scientiarium Polonorum Formatio Cirtumiectus,* **6**, 3-16.

[45] Arnold, J.G., Srinivasan, R., Muttiah, R.S. and Williams, J.R. (1998) Area Hydrologic Modeling and Assessment Part I: Model Development. *Journal of the American Water Resources Association,* **34**, 73-89. http://dx.doi.org/10.1111/j.1752-1688.1998.tb05961.x

[46] Marion, D.A., Sun, G., Caldwell, P.V., Chelcy, F., Conrads, P.A., Laird, S.G., Dai, Z., Meyers, J.A.M. and Trettin, C. (2013) Managing Forest Water Quantity and Quality under Climate Change. In: Vose, J. and Kleipzig, K., Eds., *Climate Change Adaptation and Mitigation Management Options: A Guide for Natural Resource Managers in Southern Forest Ecosystems,* CRC Press, Boca Raton, 496. http://dx.doi.org/10.1201/b15613-10

[47] Feaster, T.D., A.J., G. and Weaver, J.C. (2009) Magnitude and Frequency of Rural Floods in the Southeastern United States, 2006. Scientific Investigations Report 2009-5156, US Geological Survey, 226.

[48] Benson, M.A. and Carter, R.W. (1973) A National Study of the Streamflow Data-Collection Program. Water-Supply Paper 2208, US Geological Survey, 44.

[49] Richter, B.D., Baumgartner, J.V., Powell, J. and Braun, D.P. (1996) A Method for Assessing Hydrologic Alteration within Ecosystems. *Conservation Biology,* **10**, 1163-1174. http://dx.doi.org/10.1046/j.1523-1739.1996.10041163.x

[50] Amatya, D.M., Harrison, C.A. and Trettin, C.C. (2014) Comparison of Potential Evapotranspiration Using Three Me-

thods for a Grass Reference and a Natural Forest in Coastal Plain of South Carolina. *Proceedings of the* 2014 *South Carolina Water Resources Conference*, Columbia.

[51]　Amatya, D.M. and Trettin, C.C. (2010) Outflow Characteristics of a Naturally Drained Forested Watershed in Coastal South Carolina. 9*th International Drainage Symposium*, Held Jointly with CIGR and CSBE/SCGAB, Quebec City.

[52]　U.S. EPA (2015) Connectivity of Streams and Wetlands to Downstream Waters: A Review and Synthesis of the Scientific Evidence (Final Report). U.S. Environmental Protection Agency, Washington, DC, EPA/600/R-14/475F. http://cfpub.epa.gov/ncea/cfm/recordisplay.cfm?deid=296414

[53]　Schot, P. and Winter, T. (2006) Groundwater-Surface Water Interactions in Wetlands for Integrated Water Resources Management. *Journal of Hydrology*, **30**, 261-263. http://dx.doi.org/10.1016/j.jhydrol.2005.07.021

[54]　Colquhoun, D.J., Gardner, R.L. and Steele, K.B. (1986) Recharge-Discharge Area, Piezometric Surface and Water Chemistry Characteristics of the Tertiary Limestone Aquifer System in South Carolina. Water Resources Research Institute, Clemson.

[55]　Meadows, J.K. (1987) Ground-Water Conditions in the Santee Limestone and Black Mingo Formation near Moncks Corner, Berkeley County, South Carolina. WRC-156, South Carolina Water Resources Commission, Columbia, 38.

[56]　Bouwer, H. (1989) Bouwer and Rice Slug Test—An Update: Bouwer, H Ground WaterV27, N3, May-June 1989, P304-309. *International Journal of Rock Mechanics and Mining Sciences & Geomechanics Abstracts*, **26**, 305. http://dx.doi.org/10.1016/0148-9062(89)91577-5

[57]　Bouwer, H. and Rice, R.C. (1976) A Slug Test for Determining Hydraulic Conductivity of Unconfined Aquifers with Completely or Partially Penetrating Wells. *Water Resources Research*, **12**, 423-428. http://dx.doi.org/10.1029/WR012i003p00423

[58]　Aucott, W.R. and Speiran, G.K. (1985) Potentiometric Surfaces of November 1982 and Declines in the Potentiometric Surfaces between the Period Prior to Development and November 1982 for the Coastal Plain Aquifers of South Carolina. Water-Resources Investigations Report 84-4215, US Geological Survey.

[59]　Newcome Jr., R. (1989) Ground-Water Resources of South Carolina's Coastal Plain. WRC-167, South Carolina Water Resources Commission, Columbia.

[60]　Weems, R.E. and Lemon, E.M. (1993) Geology of the Cainhoy, Charleston, Fort Moultrie, and North Charleston Quadrangles, Charleston and Berkeley Counties, South Carolina. Misc. Invest. Ser. Map I-1935, US Geological Survey, Reston.

[61]　Plewa, T. and Hansen, W. (2003) Current Status of Water Quality Information on the Francis Marion National Forest. USDA Forest Service, Columbia, 43.

[62]　Muwamba, A., Amatya, D.M., Trettin, C. and Glover, J. (2015) Comparing Nutrient Exports from 1st, 2nd, and 3rd Order Watersheds on South Carolina Atlantic Coastal Plain. *Proceedings of the 5th Interagency Conference on Research in the Watersheds*, North Charleston.

[63]　Bureau, U.S.C. (2010) Census of Population, State & County Quick Facts: Berkeley County, South Carolina. Washington DC.

[64]　USFS (2015) Francis Marion National Forest Plan Revision Environmental Impact Statement. USDA Forest Service (under Internal Review before Public Release).

[65]　Dai, Z., Li, C., Trettin, C.C., Sun, G., Amatya, D.M. and Li, H. (2010) Sensitivity of Stream Flow and Water Table Depth to Potential Climatic Variability in a Coastal Forested Watershed. *Journal of the American Water Resources Association*, **46**, 1036-1048. http://dx.doi.org/10.1111/j.1752-1688.2010.00474.x

[66]　Miwa, M., Gardner, D.L., Bunton, C.S., Humphreys, R. and Trettin, C.C. (2003) Characterization of Headwater Stream Hydrology in the Southeastern Lower Coastal Plain. USDA Forest Service, Charleston.

[67]　Griffin, M.P., Callahan, T.J., Vulava, V.M. and Williams, T.M. (2014) Storm-Event Flow Pathways in Lower Coastal Plain Forested Watersheds of the Southeastern United States. *Water Resources Research*, **50**, 8265-8280. http://dx.doi.org/10.1002/2014WR015941

[68]　Garrett, C.G., Vulava, V.M., Callahan, T.J. and Jones, M.L. (2012) Groundwater-Surface Water Interactions in a Lowland Watershed: Source Contribution to Stream Flow. *Hydrological Processes*, **26**, 3195-3206. http://dx.doi.org/10.1002/hyp.8257

[69]　Griffin, M.P., Epps, T.H., Callahan, T.J., Vulava, V.M. and Hitchock, D.R. (2012) Using Water Chemistry Data to Assess Stormwater Pathways in Lowland Watersheds. *Proceedings of the* 2012 *South Carolina Water Resources Conference*, Columbia.

[70]　Amoah, J.K.O. (2008) A New Methodology for Estimating Watershed-Scale Depression Storage. PhD Dissertation, Florida A&M University, Tallahassee.

[71]　Amoah, J.K.O., Amatya, D.M. and Nnaji, S. (2013) Quantifying Watershed Surface Depression Storage: Determina-

tion and Application in a Hydrologic Model. *Hydrological Processes*, **27**, 2401-2413. http://dx.doi.org/10.1002/hyp.9364

[72] Dai, Z., Li, C., Trettin, C., Sun, G., Amatya, D. and Li, H. (2010) Bi-Criteria Evaluation of the Mike She Model for a Forested Watershed on the South Carolina Coastal Plain. *Hydrology and Earth System Sciences*, **14**, 1033-1046. http://dx.doi.org/10.5194/hess-14-1033-2010

[73] Amatya, D.M., Ssegane, H., Harrison, C.A. and Trettin, C.C. (2015b) Testing Resiliency of Hydrologic Dynamics of a Paired Forested Watershed after a Hurricane in Atlantic Coastal Plain.

[74] Dai, Z., Li, C., Trettin, C.C., Sun, G., Amatya, D.M. and Li, H. (2011) Climate Variability and Its Impact on Forest Hydrology on South Carolina Coastal Plain, USA. *Atmosphere*, **2**, 330-357. http://dx.doi.org/10.3390/atmos2030330

[75] Jayakaran, A.D., Williams, T.M., Ssegane, H., Amatya, D.M., Song, B. and Trettin, C.C. (2014) Hurricane Impacts on a Pair of Coastal Forested Watersheds: Implications of Selective Hurricane Damage to Forest Structure and Stream-flow Dynamics. *Hydrology and Earth System Sciences*, **18**, 1151-1164. http://dx.doi.org/10.5194/hess-18-1151-2014

[76] Wilson, L., Amatya, D.M., Callahan, T.J. and Trettin, C.C. (2006) Hurricane Impact on Stream Flow and Nutrient Exports for a First Order Forested Watershed of the Lower Coastal Plain, South Carolina. *Proceedings of the* 2nd *International Conference on Research on Watersheds*, Coweeta.

[77] NCDWQ (2005) Identification Methods for the Origins of Intermittent and Perennial Streams, Version 3.1. North Carolina Division of Water Quality, Raleigh.

[78] Trettin, C.C., Amatya, D.M., Kaufman, C., Levine, N. and Morgan, R.T. (2008) Recognizing Change in Hydrologic Functions and Pathways Due to Historical Agricultural Use—Implications to Hydrologic Assessment and Modeling. *Proceedings of the* 3rd *International Conference on Research in the Watersheds*, Aspen.

[79] Czwartacki, B.J., Trettin, C.C. and Callahan, T.J. (2015) Water Dynamics along the Terrestrial Boundary of a Tidal Freshwater Forested Wetland. Wetlands (Submitted for Review December 2014).

[80] USFS (2003) Roads Analysis: Francis Marion National Forest. USDA Forest Service, Columbia, 36.

[81] McLaughlin, K., Ragus, J. and Hansen, W.F. (2002) Soil and Water Conservation Practices Guide. USDA Forest Service, Atlanta, 149.

[82] Amatya, D., Harrison, C. and Trettin, C. (2007) Water Quality of Two First Order Forested Watersheds in Coastal South Carolina. In: *Watershed Management to Meet Water Quality Standards and Total Maximum Daily Load*, *Proceedings of the ASABE Annual Conference*, San Antonio.

[83] Amatya, D.M., Callahan, T.J., Radecki-Pawlik, A., Drewes, P., Trettin, C. and Hansen, W.F. (2008) Hydrologic and Water Quality Monitoring on Turkey Creek Watershed, Francis Marion National Forest, South Carolina. *Proceedings of the* 2008 *South Carolina Water Resources Conference*, Charleston.

[84] Wu, K. and Xu, Y.J. (2006) Evaluation of the Applicability of the Swat Model for Coastal Watersheds in Southeastern Louisiana. *Journal of the American Water Resources Association*, **42**, 1247-1260. http://dx.doi.org/10.1111/j.1752-1688.2006.tb05610.x

[85] Arnold, J.G. and Srinivasan, R. (1998) A Continuous Catchment-Scale Erosion Model. In: Boardman, J. and Favis-Mortlock, D., Eds., *Modelling Soil Erosion by Water*, Springer Berlin, Heidelberg, 413-427. http://dx.doi.org/10.1007/978-3-642-58913-3_31

[86] Miller, S.N., Kepner, W.G., Mehaffey, M.H., Hernandez, M., Miller, R.C., Goodrich, D.C., Kim Devonald, K., Heggem, D.T. and Miller, W.P. (2002) Integrating Landscape Assessment and Hydrologic Modeling for Land Cover Chane Analysis. *Journal of the American Water Resources Association*, **38**, 915-929. http://dx.doi.org/10.1111/j.1752-1688.2002.tb05534.x

[87] Santhi, C., Arnold, J.G., Williams, J.R., Dugas, W.A. and Hauck, L. (2002) Application of a Watershed Model to Evaluate Management Effects on Point and Non-Point Source Pollution. *Transactions of the ASAE*, **44**, 1559-1570.

[88] Fohrer, N., Eckhardt, K. and Haverkamp, S. (1999) Applying the Swat Model as a Decision Support Tool for Land Use Concepts in Peripheral Regions in Germany. *Proceedings of the* 10th *International Soil Conservation Organization Meeting*, West Lafayette, 994-999.

[89] Sophocleous, M.A., Koelliker, J.K., Govindaraju, R.S., Birdie, T., Ramireddygari, S.R. and Perkins, S.P. (1999) Integrated Numerical Modeling for Basin-Wide Water Management: The Case of the Rattlesnake Creek Basin in South-Central Kansas. *Journal of Hydrology*, **214**, 179-196. http://dx.doi.org/10.1016/S0022-1694(98)00289-3

[90] Amatya, D.M., Trettin, C.C., Skaggs, R.W., Burke, M.K., Callahan, T.J., Sun, G., Nettles, J.E. and Parsons, J.E. (2005) Five Hydrologic Studies Conducted by or in Cooperation with the Center for Forested Wetlands Research. SRS-40, USDA Forest Service, Asheville, 24.

[91] Lam, Q.D., Schmalz, B. and Fohrer, N. (2010) Modelling Point and Diffuse Source Pollution of Nitrate in a Rural Lowland Catchment Using the Swat Model. *Agricultural Water Management*, **97**, 317-325.

http://dx.doi.org/10.1016/j.agwat.2009.10.004

[92] Krause, C. and Roghair, C. (2010) Fish Inventory Results for Coastal Plain Streams on the Francis Marion National Forest, South Carolina. USDA Forest Service, Blacksburg, 28.

[93] Hawkins, C.P., Olson, J.R. and Hill, R.A. (2010) The Reference Condition: Predicting Benchmarks for Ecological and Water-Quality Assessments. *Journal of the North American Benthological Society*, **29**, 312-343. http://dx.doi.org/10.1899/09-092.1

9

Assessment of the Effectiveness of Watershed Management Intervention in Chena Woreda, Kaffa Zone, Southwestern Ethiopia

Yericho Berhanu Meshesha[1]*, **Belay Simane Birhanu[2]**

[1]School of Natural Resources and Environmental Studies, Hawassa University, Hawassa, Ethiopia
[2]College of Development Studies, Addis Ababa University, Addis Ababa, Ethiopia
Email: *yerichob@gmail.com

Abstract

The main purpose of this study was to assess the effectiveness of watershed management intervention in Chena Woreda. A systematic sampling technique was used to select sample micro-watersheds, and random sampling method was used to select individual households from both intervention and non-intervention areas. Data were collected through field observation, household questionnaire survey, focused group discussion, in-depth interview and key informant interview. Moreover, physical soil and water conservation structures' layout measurement was conducted. Descriptive statistics, t-test, chi-square test and participation index were used for data analyses. The study revealed that the intervention has good achievements in reducing soil erosion, improving water availability and quality, developing tree plantation and diversifying household income sources in the catchment. However, poor community participation, lack of the structures design alignment with standards, inappropriate time of implementation, lack of diversified soil water conservation measures, absence of regular maintenance and management of the structures were some of the major limitation of the intervention. Therefore, this study recommends that the stakeholders should make appropriate correction measures for observed failures and further interdisciplinary study should be conducted to explore the problems.

Keywords

Chena Woreda, Effectiveness, Intervention, Standards, Watershed Management

---

*Corresponding author.

1. Introduction

Accelerated soil erosion is the major threat to agricultural production in Ethiopia. It is estimated that soil of about 1.5 billion tons which has the monetary value of US $1 to 2 billion per year is being eroded every year. The rate of erosion in highlands of the country is extreme and reaches up to 300 tons per hectare annually [1] [2]. Out of 60 million hectares of estimated agriculturally productive land: 27 million hectares are significantly eroded, 14 million hectares are seriously eroded and 2 million hectares reach at the point which is irreversible [1]. Deforestation and land degradation through accelerated soil erosion in turn result in low productivity and poverty in the country [3].

Moreover, for the last several decades, different human activities such as developments in controlling and diverting surface waters, exploring ground water, overgrazing and over use of natural resources for a variety of purposes have been undertaken without care. Absence of conserving the natural resource, mismanagement of the watershed, and lack of preserving the quality of environment have greatly impaired the sustainable development of the country. The consequences include deforestation, land degradation, water shortage, pollution, flooding, impaired fisheries, and reduced recreational opportunities. The findings of several studies in the country reveal that the uses of natural resource are exceeding the carrying capacity of the ecosystem. For instance, in Awash valley wetlands were dried up due to irrigation development projects; pastoral lands were transformed into cotton production and extensive wetlands drainage resulted in drying up of 150 springs in Illubabor Metu Woreda [4].

In order to alleviate the aforementioned problems, the role of effective watershed management is indispensable. It can prevent community water shortage, poor water quality, flooding and erosion. Consequently, the rivers, streams, wetlands and lakes of a given watershed area can provide ecological services that maintain the health, safety, economy and social welfare by storing and purifying drinking water, providing recreational opportunities that attract tourists, maintaining biological diversity, providing spawning opportunities for commercially valuable fish, raising property values, supporting agriculture, and commencing and protecting people and property from flooding [5].

In Ethiopia, there was no governmental policy on soil and water conservation and natural resource management prior to 1974. The 1974-1975 famine has made the turning point for the country to conserve her natural resources [6]. Watershed development planning with the aim of natural resource conservation and development programs was started in the 1970s [7] [8]. A large scale effort has made to implement the watershed management projects in the country. However, due to its large scale planning units which range 30 to 40 thousands of hectares and absence of local community participation, the projects were ended with unsatisfactory results. The failure of large-scale watershed management projects had opened an opportunity to the stakeholder's especially the researchers, governmental organizations and NGOs to find out solution for the problem. Hence, FAO in its pilot study from 1988 to 1991 found that watershed management approach which focuses on a bottom-up basis and uses smaller units is vital to attain the overall designed watershed management objective.

During the last two decades, MoA, GTZ, FAO and SOS Sahel have adopted participatory land use planning in different parts of the country. The interventions in South Gonder, North and West Shoa of Oromia, some parts of Tigray, North Wello and Wolaita were implemented by technical support of NGOs under Ministry of Agriculture through participatory basis. Since 2005 the country has developed community-based watershed development guideline, in which the participation of community gets due consideration for sustainable watershed development and management [8]. Recently, the movement on watershed management is going on almost throughout the country. Besides the efforts made by several NGOs, the campaign on soil and water conservation program which was initiated by FDRE government for the last four years has offered a great contribution in watershed development and management for the country.

In spite of having the aforementioned efforts on watershed management development, the effectiveness of the intervention was not often evaluated. The experience in Ethiopia showed that the practice of evaluation of the effectiveness of the project is overlooked for three main reasons such as lack of political administrative commitment, insufficiency of budget allocated for monitoring and evaluation and inadequacy of the institutional arrangements that underlie monitoring and evaluation [9].

Therefore, this study was designed to evaluate the effectiveness of watershed management intervention in Chena Woreda. Chena Woreda has started watershed management intervention in collaboration with FAO since 2008. The watershed management project has been implemented in 13 sub-districts out of 42 sub-districts in the Woreda [10]. Effective watershed management is the one that has achieved the planned objectives of the intervention [11]. Therefore, this paper conceptualized that watershed management intervention that is designed and managed to maintain the function of environmentally friend, economically viable, socially acceptable, and institutionally and technically sound is effective.

2. Materials and Methods

2.1. Study Area Description

The study was conducted in Chena Woreda which is geographically located between 7° up to 7°45'N latitude and 35°69' up to 36°06'E longitude. Chena Woreda is situated in Kaffa Zone of the Southern Nations, Nationalities and people's region. It is about 510 km away from Addis Ababa: the capital city of Ethiopia to Southwestern on the main road of Jimma to Mizan Teferi, and it is about 70 km from the zonal town of Bonga [12] (**Figure 1**).

The total area of Chena Woreda is estimated to be 901.92 km^2 [12] [13]. The total population of Woreda was 183,335: consisting of 90,400 men and 92,935 women in [14]. The Woreda receives rainfall almost all the year round [13]. The average annual minimum and maximum rainfall is 1379 mm and 1889 mm respectively. The mean monthly temperature ranges from 14°C to 28°C. The altitude of the Woreda ranges from 1000 to 3000 m.a.s.l [12].

2.2. Methodology

The study is conducted on four sub watersheds which are selected systematically: two have watershed management intervention, while the rests haven't intervention. Accordingly, *Wota-Wora* and *Woda-Kulish* from intervention, while *Dosha-Kosa* and *Boba-Bela*micro watersheds from non-intervention were selected considering their historical similarity before watershed management intervention. This systematic selection was done in order to use best matched watersheds for comparison.

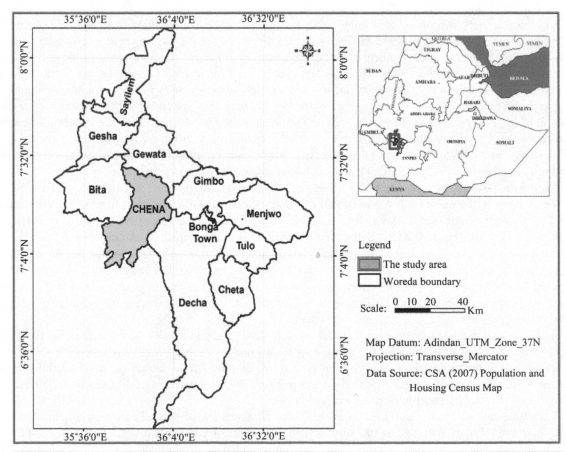

Figure 1. Map of the study area, central statistics agency.

On average each micro watershed has 500 households [10]. Hence, the selected four sample micro watershed has total of 2000 households. Taking sample size of 10% of population for the total population more than 1500 is sufficient [15]. Therefore, 10% of households which was 200 households; proportionally 50 households from each sample micro watershed were used for the questionnaire survey. The selection of individual household from all sample micro watersheds was done through simple sampling technique. Before beginning the actual work, permission was requested from the local administrations to carry out the research. Formal and informal discussions with leader of peasant association, district institutions, and villagers were conducted. Based on the information obtained from the discussions data collection process was employed. The selected household member either male or female who has age above 18 was used for household questionnaire survey. Besides to this survey, 8 focus group discussion and 12 key informants interview was conducted. The focused discussion has incorporated 6 to 12 people in each group. Community elders, youth and females were included in the focused group discussion.

Field observation was focused on observation of biophysical characteristics of Watershed like land degradation, crop patterns, distribution of settlements, individual activities in the farming plots, farmers' land management practices, water resources, bush and grazing lands, and other relevant aspects in the catchment. The observation was covered all sampled micro watershed in the study area. During this field observation river course characteristics including water quality, availability, color and odor of water, water source protection systems or mechanisms and ecological conditions were observed.

Household Questionnaire Survey was used to collect the primary data from sample households. The survey was conducted by using both open and closed ended structured questions. It was focused on individual household's characteristics on both intervention of watershed management and nonintervention. Also it was focused to get information on farmer's field practices of land resource management.

Focused group discussion was conducted based on checklists and semi-structured questionnaires, and in-depth interview was used for collection of data. During this session, respondents expressed their opinions, views, feelings and perspectives about the project process and outcomes. Moreover, key informants interview was carried out with 4 elders, 4 local administrators and 4 experts.

Soil and Water Conservation structures layout measurement was conducted on sample households' field in the intervention area. At least one SWC structure measurement was taken in their type on individual sample households' field. After completion of editing the process of assigning numerical symbols (coding) to answers were done and then the collected data were entered in to Statistical Package for Social Science, version 20. Then, descriptive statistics, t-test, chi-square test, participation index and logistic regression model were used for analysis.

Descriptive statistics such as frequency of information, percentage, mean and standard error of mean was used. Frequency of information was mainly used to analyze categorical qualitative data. The independent and one sample t-test was used. Independent t-test was used to compare the means of the parameters in intervention and nonintervention areas. One sample t-test was used to compare the observed means of SWC structures layout with the standards. Chi-square test was also used to compare significance the mean variation between the two or more groups of the categorical variables.

Participation index was used to analyze the participation of individual respondents and the community. Determination of individual respondent's participation was done by using the following equation (Equation (1))

$$PI_i = \frac{\sum_{j=1}^{A} Y_{ji}}{A}$$　　　　(1)

where PI_i = Participation index of the ith respondent, Y_j = 1, if the respondent has participated in the jth activity, Y_j = 0, if the respondent has not participated in the jth activity and A = Total number of Activity. Determination of the participation status of the community in the catchment is done using the following equation (Equation (2)).

$$PI_c = \frac{\sum_{i=1}^{N} PI_i}{N}$$　　　　(2)

where, PI_c = Participation index for the community, PI_i = Participation index of individual respondent and N = Total number of respondents.

3. Results and Discussion

3.1. Effects of Intervention on Soil Fertility

Based on a diversified criteria and knowledge, farmers evaluated the fertility of soil in their own land in the catchments. Conventionally, farmers in the study area categorize their land in to three fertility status namely: good, medium and poor.

The survey result depicted that majority of the respondents in both sites categorized their land in to moderate and good fertility classes, and relatively high percentage of respondents in the intervention area responded that their soil fertility as good. The test statistics result (P > 0.05) shows that observed differences was not statistically significant (**Table 1**). This result indicates that the watershed management intervention didn't bring improvement of soil fertility in the area. This finding is contradicts with the findings of [16] and [17], and the general principles of watershed management [18]. This is because of that having several locally used soil fertility indicators; farmers mainly associate the status of soil fertility with the productivity of land.

3.2. Trends of Soil Erosion

The farmers' perception on the trends of soil erosion is manifested by stating their view as erosion is increasing, decreasing or as no change.

As shown in **Table 2**, 58% respondents in intervention area responded that the rate of soil erosion which has great impact on soil fertility has been decreasing since 2008. On the other hand, all respondents in the non-intervention area respond that the trend of soil erosion is increasing from time to time in their respective micro watershed. The statistical test result presented in **Table 2** depicted that there was significant variation in the responses of the respondents, and it showed that the intervention has significantly reduced the soil erosion. Moreover, the information obtained from focused group discussion and key informants interview showed that the watershed management intervention reduced the rate of soil erosion as compared before intervention.

3.3. Effects of Intervention on Water Availability and Quality

The local communities categorized the availability of water into three classes: namely, good, moderate and poor. The presence of water sources, the volume of water sources, its nearness to the settlement and constant flow rate of the water sources were the main criteria locally used to categorize the availability of water resources in the

Table 1. Distribution (%) of respondents rating of soil fertility.

Soil fertility category	Site		X^2	P-value
	Intervention	Non-intervention		
Good	44	36	0.80	0.371
Moderate	39	46	0.576	0.448
Poor	17	18	0.029	0.866
n	100	100		

Table 2. Perception of farmers' on the trends of soil erosion in the area.

Trends of erosion	Response (%)		X^2	P-value
	Intervention	Non-intervention		
Increasing	3	100*	91.35	0.00
Decreasing	58*	0	-	
No change	39	0	-	
n	100	100		

* is significant at 95% of confidence.

area. According to respondents, the area which has several water sources with continuous flow rate and nearest to the settlement is characterized as good. On the other hand, if the area has no several water sources, fluctuated flow rate and far away from the settlement, the water availability is considered as poor. Moreover, if the case is in between the above two category (good and poor), the water availability is termed as moderate.

Majority of respondent in the intervention site rated water availability as good, while non intervention site as moderate. The observed result shows statistically significant variation in water availability in both areas. The availability of water in the intervention area is better than the nonintervention one (**Table 3**). Moreover, all key informants from the intervention micro watershed also expressed that the SWC structures constructed on farm land have contributed to the water from rainfall to be enter in the soil rather than being runoff, and it enhanced the soil moisture content. These indicate that the intervention has positive contribution for the improvement of water availability in the area. This result is similar with findings of [19] in India. Moreover, this result is consistent with findings of [16], in which he found out that the support of watershed management project has contributed for potential water sources availability in the catchment.

As usual, the communities classify the water quality broadly in to three classes, namely goo, moderate and poor. The basic criterion used is the turbidity of water. Accordingly to this system, high turbidity water is poor, while low turbidity is good quality.

The majority of respondents (63%) from intervention micro watershed rated the water quality as good (**Table 4**). This result is significantly higher than the others, and it shows that the watershed management intervention made improvements on water quality. Moreover, the FGD and Key informants interview in intervention site were described that water quality has been improving due to the intervention. They directly relate the reduction of erosion with water quality. It means that implementation of any measure which can reduce erosion contributes to maintain water quality through hindering the entrance of runoff in to water. This finding has similar implication with the findings of [19] and [20]. They found out that the improvement in quality of ground and surface water were observed in Indian watersheds within short period of time due to watershed development program.

3.4. Effects of Intervention on Tree Planting Practice

The household survey result showed that the practice of regularly planting of trees annually was adopted by 60%

Table 3. Distribution of respondents rating of water availability.

Water availability	Response (%)		X^2	P-value
	Intervention	Non-intervention		
Good	64*	22	20.510	0.000
Moderate	29	72*	18.307	0.000
Poor	7	6	0.077	0.752
n	100	100		

* is significant at 95% of confidence.

Table 4. Distribution of respondents rating of water availability.

Water quality	Response (%)		X^2	P-value
	Intervention	Non-intervention		
Good	63*	41	4.654	0.031
Moderate	32	46	2.513	0.113
Poor	5	13	3.556	0.059
n	100	100		

* is significant at 95% of confidence.

and 33% of respondents in intervention and non-intervention area respectively. The statistical test result presented in **Table 5** shows that there was significant difference in tree planting practices between the two sites. There was higher proportion of community members in the intervention area who were regularly planting trees than the non-intervention one. This result is in line with the findings of [16] in Tsegur and Tsegur Eyesus watershed, in which higher proportion of households were engaged in tree planting activities. Moreover, this finding is consistent with the findings of [21] in Kalu Woreda. This significant difference is because of that the support of public nursery was restricted to the intervention area. As it was observed during field visit, there was at least one public nursery site in intervention micro watershed whereas there was no public nursery site in nonintervention sites. Moreover, 83% of respondents from intervention site reported that the watershed management program has initiated the farmers to plant tree on their land through creating awareness on the importance of trees and providing tree seedlings.

3.5. Effects of Intervention on Major Household Income Sources

Agriculture is the main income source of the community in Chena Woreda. Mixed farming which involves crop production and animal husbandry is adopted by all farmers. Crop production in the area includes the production of staple food crops, cash crops and cereal crops. The production of staple food crop is limited to homestead and household consumption. Coffee production is one of well known cash crops practiced by some framers. Cereal crops such as *Zea Mays*, *Eragrostis tef*, *Vicia bean*, *Sorghum bicolor* and *Triticum* were produced by all farmers in the area.

The independent t-test result in **Table 6** depicted that the production of *Zea mays* and *Eragrostis tef* has no significant difference in both areas, while *Sorghum bicolor*, *Vicia bean* and *Triticum* in intervention area was significantly higher than in nonintervention area. In contrast with this result the FGD, key informants interview result shows the tendency of decline of crop production in the area, since farmers consider *Zea mays* as the major cereal crop. The focused group discussion participants proposed that the proportion of land used for maize production covers three fourth (75%) of the households total land size. Since, maize is the most widely produced cereal crop; the insignificant difference on this crop result indicates that the intervention did not brought improvement on crop production. This finding is argued with the findings of [21]. His study has found out that the watershed management project has contributed through maximizing the amount of food production.

Table 5. Frequency of respondents' tree planting practice.

Regular tree planting	Site		X^2	P-value
	Intervention	Non-intervention		
Yes	60*	33	7.839	0.005
No	40	67*	6.813	0.009
N	100	100		

* is significant at 95% of confidence.

Table 6. Cereal crop production (Kg/ha/year).

Crops	Intervention		Non-intervention		T-value	P-value
	Mean	SE	Mean	SE		
Zea mays	835	61.42	797	35.78	0.529	0.264
Eragrostis tef	315.5	23.73	282	14.13	1.213	0.114
Sorghum bicolor	937	79.8	677	42.99	1.732*	0.043
Vicia bean	677.6	56.0	557.5	25.98	1.912*	0.029
Triticum	634.5	46.4	666.6	21.94	2.429*	0.008
n	100		100			

* is significant at 95% of confidence.

The production of animal husbandry is also well practiced by all farmers in the area. Cattle, sheep, goats, poultry, horse, mule and donkey were among animals raised for both source of food and commercial purposes.

As presented in **Table 7**, the total livestock unit (TLU) of the intervention area was higher than the nonintervention one. As all focused group participants and key informants have mentioned, the intervention has contributed to livestock management by providing fodder seedlings. This result is consistent with findings of [16], he found out that the availability of fodder due to the intervention has contributed for better livestock production in Dijjil and Lenche-Dima catchments.

3.6. Community Participation in Watershed Management Intervention

The participation index calculation results of the respondents were presented in **Table 8**.

As shown in **Table 8**, 17% of the respondents didn't participate in any activities of the watershed management intervention. Moreover, all respondents participated in less than 60% of the intervention activities. The overall participation index of the community in the watershed management was 0.42. This shows that the community did not participate in the majority (58%) of project activities, which is poor participation. The result of this study is similar to the findings of [19] in India and [22] in Chemoga watershed of Amhara regional state. They found out that the community participation in the watershed management was poor. During focused group discussion, the participants raised several complains in watershed management and community participation in their respective catchment. They complained that the strategy of watershed management for the area has come from the donor of the project and the communities didn't participate right from beginning. Due to this reason, it was taken several years for the project to create awareness rather than implementing the program. Moreover, the community participation in project evaluation and monitoring wasn't adopted. Due to this reason, the overall participation status of the community is less than the average (50%).

The active involvement of women in watershed management is vital for effectiveness of watershed management. The assessment results of this study on women participation in watershed management were presented in **Table 9**.

The result presented in **Table 9** shows that the overall participation of women is limited to only 20.6% of watershed management activities. This means that the participation status of women in watershed management of the area is inadequate. The findings of this study is similar with [19] and [22] findings. The key informants and

Table 7. Livestock production by of Total Livestock Unit (TLU)

Livestock	Intervention area	Non-intervention
Cattle	22,920	14,876
Sheep	652.21	511.94
Goat	588	410
Poultry	163.57	115.75
Horse	1463	401.5
Donkey	59.5	25.9
Mule	919.6	163.9
Total	26,765	16,505

Table 8. Participation index (PI_i) of the individual respondent.

PI_i	Frequency	Cumulative percent	PI_c
0.00	17	17	
0.2	16	33	
0.4	7	40	0.42
0.6	60	100	

Table 9. Participation index (PI_i) of the women.

PI_i	Frequency	Cumulative percent	Overall women participation index
0.00	32	32	
0.2	33	65	0.206
0.4	35	100	

focus group discussion members stated that culturally, the task of women is working in the home rather than in the field activities. Therefore, culture implication is the main reason for low status of women participation in watershed management. Similarly, [22] has found the similar reason in Chemogo watershed.

3.7. Structural Arrangement of Institution in the Watershed Management

The structural arrangement of watershed management institution in the study area has comprised several people at different stages. The members include focal person at Woreda level, community facilitator, community leaders, watershed management team and communities (**Figure 2**).

The focal person is the responsible for watershed development and management in the Woreda. He/she has the responsibility of regulating and controlling the activities going on in the watershed management intervention area. The focal person has the duty to report the activities performed to Woreda Agriculture and Rural Development office regularly.

The community facilitator is one for each micro watershed. He/she has the responsibility of facilitating the activities in the watershed management, teaching the community for watershed development, and serve as foreman during structure implementation. The facilitator has the responsibility of reporting the activities to the focal person monthly. Community facilitators were employed by the project. However, they were not fully devoted since the employment is on the bases of contract agreement.

Community Leaders are administrative bodies of the catchment including manager, chairman, secretary and members. They have the responsibility of regulating and coordinating watershed management activities in coordination with community facilitators and experts. However, most respondents of the intervention area (71 out of 100 which is 71%) said that the local government representatives (community leaders) were doing to secure their authority for long time rather than improving the communities' livelihood. The local administrative body forces the community to accept intervention programs without their own interest.

The teams are the community members who are elected by the community and approved by community leaders. Each committee contains seven members; one leader, one vice, one secretary, one youth representative, one female representative and two members.

The structural arrangement of the watershed management of the area seems to be good because it was planned to include the members of the communities as its main components. However, it has several limitations. Firstly, the assigning of focal person wasn't considered technical skill, knowledge and educational background of the person. Secondly, the contract agreement between Woreda Agriculture and Rural Development and community facilitators has made the facilitators to hesitate to do the work with their maximum effort since it is only for short period of time. The in-depth interview conducted with representatives of Woreda sectors showed that the coordination of watershed management sector with other sectors is poor. In addition, the Agriculture and Rural Development Office representatives stated that the coordination they do have with other sector was also meager.

3.8. Technical Viability of Implemented SWC Structures

As the field survey result depicted, Level fanya juu, Level soil bunds, Level stone fenced soil bunds and stone bunds were implemented SWC structures in the area. Through considering the agro-ecology and soil depths in the area, the Level fanya juu and Level soil bund are suitable for the area, while level stone fenced soil bunds and stone bunds are not appropriate for the area.

Evaluation of the fitness of Level fanya juu to the standard layout of the structure was done based on the data obtained from field measurement (**Table 10**).

The field measurement results of physical layouts of the structures presented in **Table 10** shows that only Top width of embankment is in line with standards. Therefore, the results presented in **Table 10** clearly indicated

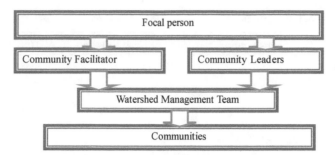

Figure 2. Structural arrangement of watershed management intervention in the area.

Table 10. Observed layouts of Level fanya juu.

Variables	Mean (m)	SD	Test Value (m)	MD (m)	T-Value	P-Value
Length of Level Fanya Juu	9.44	0.77	10	−0.556	−5.653*	0.000
Width of the Ditch	0.46	10.16	0.5	−0.044	−3.649*	0.001
Depth of the Ditch	0.41	10.14	0.5	−0.95	−7.31*	0.000
Length of Tie Ridge	0.41	10.77	0.5	−0.086	−6.243*	0.000
Height of Embankment	0.34	18.30	0.5	−0.163	−6.961*	0.000
Length of Berm	0.37	14.47	0.25	0.116	6.275*	0.000
Top Width of Embankment	0.32	0.103	0.3	0.023	1.781	0.08
Bottom Width of Embankment	0.81	15.54	1.6	−0.793	−39.86*	0.000
Vertical Interval	1.91	0.41	1.5	0.41	7.842*	0.000
Valid n			61			

*Significant at 95% confidence.

that the implemented physical layouts of Level fanya juu weren't in accordance with standards.

The one sample t-test result presented in **Table 11** indicated that the majority of variables of Level soil bund layouts such as length of Bund, depth of Ditch, length of Tie ridge, height of Embankment, length of Berm (the distance between the mouth of Ditch and Embankment) and bottom width of embankment were significantly lower than the standards. On the other hand, the observed mean Vertical interval was significantly higher than the standard. However, the top width of the Embankment of Level soil bund was only constructed with the standard.

The key informants stated that these structural layout faults come from two main sources. Firstly, some farmers' were bargaining the foremen during construction of the structure because they assume that nearly constructed structures occupy their land which is useful for cultivation. Secondly, there is lack of skills. The selected foremen were the member of community and they did not get detail training on technical issues of SWC structures. Thirdly, the undulating topography of the area doesn't allow constructing the structures as design. In spite of having the failure to meet the standards, FGD and key informants interview shows, the implemented SWC structures has reduced the rate erosion in the area.

3.9. Diversity of SWC Structures

One cannot fit the entire syndrome. Hence, no one structure is totally suitable for given area. Majority (66%) of the respondents revealed that one structure type per farm land plot of individual farmer was implemented, and 23 % of respondents were used two types of structure per farm land plot. The remaining one percent used three and more structure per farm plot. These figures indicated that the diversity of structure per farm land plot of each farmer land is very poor. Moreover, the diversity of physical SWC structure at Woreda level is also poor since only the aforementioned four structures were dominated in the Woreda.

Table 11. Observed layouts of Level soil bund.

Variables	Mean (m)	SD	Test Value (m)	MD (m)	T-Value	P-Value
Length of Level Fanya Juu	9.42	0.79	10	−0.548	−4.49*	0.000
Width of the Ditch	0.45	0.12	0.5	−0.054	−2.75*	0.009
Depth of the Ditch	0.41	0.11	0.5	−0.095	−5.49*	0.000
Length of Tie Ridge	0.42	0.11	0.5	−0.084	−4.80*	0.000
Height of Embankment	0.32	0.19	0.5	−0.176	−5.73*	0.000
Length of Berm	0.38	0.15	0.25	0.127	5.27*	0.000
Top Width of Embankment	0.32	0.10	0.30	0.025	1.43	0.160
Bottom Width of Embankment	0.81	0.16	1.6	−0.79	−30.1*	0.000
Vertical Interval	1.86	0.41	1.5	0.36	5.30*	0.000
Valid n	37					

*Significant at 95% confidence.

3.10. Maintenance and Management of the Structure

Regular maintenance and management of the implemented SWC measures should be done for its sustainability in the area. However, the trends in the study area indicated that there wasn't regular maintenance and management of once implemented structures. Majority of respondents (78.6%) agreed that there was no maintenance and management activity once the structures were constructed. As the key informants' information, once the structure is constructed no one go back for its maintenance and management. The participants of all focused group discussion have also concluded that there was no regular maintenance and management of implemented structures in the area. They also confirmed that the problem wasn't only the farmers but also the Woreda is too. The Woreda agriculture and rural development has given attention mainly to the expansion of the structure. Due to this, there was limitation in regular maintenance and management of implemented structure. Hence, during field observation the destructed structures were mainly observed in the area. This finding is similar with [23], by which Kebede found poor structure maintenance in Campaign works watershed Management.

3.11. Timing of SWC Measures Implementation

The household survey result revealed that 50% of SWC measures were implemented during January, 22% from October to December and 28% from February to April. This survey result depicted that the majority of SWC measures were implemented at January. On the other hand, January is an intensive cultivation season for the Woreda. Hence, the SWC measures implementation program was overlapped with the intensive cultivation seasons in the area. In this regard, all respondents complained about the timing of SWC measure implementation. Moreover, all key informants and the Woreda Agriculture and Rural Development experts of Natural Resource Management had complained the timing of SWC measures implementation. The expert said that the improper timing of SWC measures implementation is due to the miss match of the Woreda growing season with other Woredas' of the region. The implementation calendar was come from the region (SNNPR) and therefore, the Woreda is facing the challenges in implementation program arrangements.

4. Conclusions

The watershed management intervention in Chena Woreda was effective in several aspects; meanwhile it has also the components in which the project has unsatisfactory achievements. The findings indicated that the watershed management intervention brought reduction in soil erosion and thereby improvement of quality and water availability, and development plantation forest in its intervention area. The study further disclosed that there was no significant improvement in crop production in the Woreda. However, the intervention brought an improvement in livestock production.

The participation status of communities in watershed management was poor. The structural arrangement of the institution was participatory, while the technical skill, knowledge and capacity of the assigned persons in the positions were poor. Similarly, some local administrative bodies have their own interest and motives like securing authority for long period of time rather than devoting themselves for the community development through sustainable watershed management.

Majority of introduced physical SWC structures are appropriate for the area. However, the implemented structures layouts were not in accordance with the standards. Besides to the limitation on layouts of the structures, the diversity of implemented structure was also poor. Moreover, the regular maintenance and management of implemented SWC structures were not practiced in the area. Timing of SWC structures implementation in the Woreda was also inappropriate. In spite of having the limitations due to several reasons, the overall evaluation showed that watershed management intervention has good achievements in the area.

5. Recommendations

- The local communities are expected to play their role by actively participating in natural resources conservation.
- The local government is expected to develop the skill, knowledge and capacity of focal person, community facilitators, community leaders, watershed management teams and the communities in relation to watershed management through capacity building.
- The SWC structures should be implemented in accordance with the standards. Therefore, it is advisable that the project implements structures according to the standard layouts. The intervention should use a diverse SWC measures. Regular maintenance and management of the structures should be in the place. Moreover, the appropriate time of the SWC for the local situation is expected to be identified and used for the further implementation.
- Furthermore, interdisciplinary study for better development of the project is recommended to be done in the same study area or elsewhere in the country to provide empirical evidences for the country situation.

Acknowledgements

We would like to thank the Horn of Africa Regional Environment Center and Network and Ethiopian Development Research Institute for financial support.

Endnotes

Woreda is equivalent to district.

References

[1] FAO (1984) Ethiopian Highlands Reclamation Study. Final Report, FAO, Rome.

[2] Tamirie, H. (1997) Desertification in Ethiopian Highlands, Norwegian Church AID Addis Ababa Ethiopia. RALA Report No. 200.

[3] Badege, B. (2001) Deforestation and Land Degradation in the Ethiopian Highlands: A Strategy for Physical Recovery. *Northeast African Studies*, **8**, 7-25. http://dx.doi.org/10.1353/nas.2005.0014

[4] Abebe, Y. and Geheb, K., Eds. (2003) Wetlands of Ethiopia. *Proceedings of a Seminar on the Resources and Status of in Ethiopia's Wetlands*, Addis Ababa, 2003, 49-57.

[5] Conservation Ontario (2001) The Importance of Watershed Management in Protecting Ontario Drinking Water Supplies. Ontario's Conservation Authorities, Toronto.

[6] Alemneh, D. (2003) Integrated Natural Resources Management to Enhance Food Security: The Case for Community-Based Approaches in Ethiopia. FAO, Rome.

[7] Hurni, H. (1986) Guidelines for Development Agents on Soil Conservation in Ethiopia. Community Forest and Soil Conservation Development Department, Switzerland.

[8] Lakew, D., Carucci, V., Asrat, W. and Yitayew, A., Eds. (2005) Community Based Participatory Watershed Development. Ministry of Agriculture and Rural Development, Addis Ababa, Ethiopia.

[9] Wassie, B. (2000) Project Planning, Implementation and Evaluation. United Nations Center for Regional Development, Africa Office, Kenya.

[10] CWARDO (2012) Chena Woreda Agriculture and Rural Development Annual Report of 2012. CWARDO, Wacha, Ethiopia.

[11] IFAD (2003) A Methodological Framework for Project Evaluation: Main Criteria and Key Questions for Project Evaluation. International Fund for Agriculture Development Office of Evaluation and Studies, Rome.

[12] CWFEO (2012) Statistical Abstract of the Chena Woreda, Wacha, Ethiopia.

[13] Tezera, C. (2008) Land Resources and Socio-Economic Report of Bonga, Boginda, Mankira and the Surrounding Areas in Kaffa Zone, SNNPRS, Ethiopia. Public-Private Partnerships, Addis Ababa.

[14] CSA (2011) Statistical Abstract. Federal Democratic Republic of Ethiopia Central Statistical Agency, Addis Ababa.

[15] Gay, R.L., Mills, E.G. and Airasian, P. (2009) Educational Research: Competencies for Analysis and Applications. 10th Edition, Pearson Education, Upper Saddle River.

[16] Tesfaye, H. (2011) Assessment of Sustainable Watershed Management Approach Case Study Lenche Dima, Tsegur Eyesus and Dijjil. MSc Thesis, Cornell University, New York.

[17] Kajiru, G.G., Merima, P.B., Mjbilinyi, P.B., Rwehumbiza, B.F., Hatibu, N., Mowo, G.J. and Mahoo, F.H. (2005) Assessment of Soil Fertility Status under Rainwater Harvesting on the Ndala. Sokoine University of Agriculture Press, Morogoro.

[18] Blanco-Canqui, H. and Lal, R. (2008) Principles of Soil Conservation and Management. Springer, Dordrecht.

[19] Singh, P., Behera, H.C. and Singh, A. (2010) Impact and Effectiveness of "Watershed Development Programmes" in India Review and Analysis Based on the Studies. Center for Rural Studies, Mussoorie, India.

[20] Shah, A., Devlal, R., Joshi, H., Desai, J. and Shenoy, R. (2004) Benchmark Survey for Impact Assessment of Participatory Watershed Development Projects in India. Gujarat Institute of Development Research, Ahmedabad, New Delhi.

[21] Mintesinot, A. (2007) Watershed Management: Effects and Problems, the Case of MERET Project in Kebelie-Chekorti Sub-Catchment, Kalu Woreda, Amhara Regional State. MA Thesis, Addis Ababa University, Addis Ababa.

[22] Azemeraw, A. (2010) Effectiveness and Governance of Community Based Participatory Watershed Management in Choke Mountain: The Case of Chemoga Watershed, Gojam. MA Thesis, Addis Ababa University, Addis Ababa.

[23] Kebede, W.W. (2015) Evaluating Watershed Management Activities of Campaign Work in Southern Nations, Nationalities and Peoples' Regional State of Ethiopia. *Environmental Systems Research*, **4**, 6. http://dx.doi.org/10.1186/s40068-015-0029-y

Abbreviations

CSA—Central Statistics Agency
CWARDO—Chena Woreda Agriculture and Rural Development Office
CWFEO—Chena Woreda Finance and Economy Office
FAO—Food and Agricultural Organizations of United Nation
FDRF—Federal Democratic Republic Ethiopian
GTZ—Germen Technical Cooperation
MoA—Ministry of Agriculture
SNNPR—Southern Nation Nationalities Peoples Region
SOS—Save the Children
SWC—Soil and Water Conservation
TLU—Total Livestock Unit

10

The Modern Problems of Sustainable Use and Management of Irrigated Lands on the Example of the Bukhara Region (Uzbekistan)

R. Kulmatov[1], A. Rasulov[2], D. Kulmatova[1], B. Rozilhodjaev[1], M. Groll[3]

[1]Department of Applied Ecology, National University of Uzbekistan, Tashkent, Uzbekistan
[2]Pedagogical University Named after Nizami, Tashkent, Uzbekistan
[3]Faculty of Geography, University of Marburg, Marburg, Germany
Email: rashidkulmatov46@gmail.com

Abstract

The Central Asian lowlands are characterized by an arid and continental climate. At the same time, the large streams and rivers have been providing water for the development of flourishing oases and extensive irrigated farming areas. Bukhara is one of those oases. The population of 1.7 mln. and especially the agricultural sector (with an irrigated area of 275,000 ha) use a considerable amount of water. But as the flat topography does not provide sufficient natural drainage, water logging and raising groundwater tables have become serious problems for the agricultural productivity. The combination of the high salinity of the irrigation water and the generous application of fertilizers leads to a widespread soil salinization. Excessive leaching is supposed to reduce the top soil salinity, but as the drainage system is only covering a small portion of the irrigated areas and is in need of maintenance, this process only contributes to the ongoing salinization and the reduction of soil fertility and crop yields. The data presented here for the years 2000 to 2013 indicate that the groundwater table is rising throughout the region while the groundwater salinity is decreasing. The soil salinity on the other hand is, after an improvement during the first half of the study period, slightly increasing since 2009, which also is reflected in the slight worsening of the condition of the reclaimed land during the same period.

Keywords

Water Resources, Water Quality, Irrigation Farming, Groundwater, Soil Salinization, Uzbekistan, Central Asia

1. Introduction

Most of the plains in the Aral Sea basin are characterized by a high natural soil salinity. In the floodplains the salinity is increased by the accumulation of salty minerals eroded in the upstream mountainous regions. Due to the arid climate and the intensive irrigation farming the floodplains are also prone to the hazardous development of secondary soil salinization [1] [2]. Furthermore, the Aral Sea Basin countries, especially Uzbekistan and Turkmenistan, are heavily impacted by the climate change because of the high sensitivity of the arable lands in the arid lowlands [3]-[9] as well as a strong population and economic growth and increasing demands for the food safety [10] [11]. The climate change (increase of the air temperature and the evapotranspiration), long-term reduced runoff from the Central Asian glaciers and more frequent droughts also increase the water consumption for irrigation. As a consequence of this, the soil salinity will further increase and the productivity of the agricultural lands will continue to deteriorate.

The regions that are most sensitive to these anthropogenic and climatic changes are the lower reaches of the Syr Darya, Amu Darya and Zarafshan River, which are all characterized by large scale irrigation schemes of great economic importance [10]-[12]. The agriculture is a key economic sector in Uzbekistan, currently providing about 18% of the GDP with 27% of employment in this sector. Currently 45.3% of all Uzbek agricultural land is used for the grain production (39.5% wheat), followed by cotton (36.2%), fodder crops (8.6%) and vegetables (4.7%), showing an emphasis on food security and the cash crops [13]. The irrigated lands make up only 15% of the total farmland in Uzbekistan, but provide more than 90% of all agricultural products, making the country the sixth largest cotton producer and the second largest cotton exporter in the world [10]. This economy is solely dependent on the water intense irrigation, as without the current (low efficiency) irrigation network only 10% - 20% of the Aral Sea basin population could be sustained in this mostly arid region [14]. Over the course of the last 30 to 50 years the foundation of this economy, the arable land in the arid lowlands has been subject to extensive salinization, water and wind erosion, increasing groundwater mineralization and raising groundwater tables as well as the contamination with various pollutants (heavy metals, fluorides and pesticides) [6] [12] [13] [15]-[20]. 50.1% of all irrigated lands in Uzbekistan were already affected by salinization in 1994 and due to the extensive flood irrigation and leaching more than 30% of the topsoil humus has been washed out since the 1960s [14]. Two-thirds of the arable land in Uzbekistan has today a humus content of less than 1%, which is far below the international average for agricultural land of 3% [21]. Salinization and soil depletion lead to reduced crop yields (−20% - 30% have been observed for cotton on soils with medium salinity [21]). As a result, the agricultural productivity is with 700 USD/ha at present only one-third of what it had been in the 1980s and the annual financial losses are 1 bln. USD for Uzbekistan alone [10] [22]. The loss of soil productivity is combatted by the extensive use of mineral fertilizers [23]. Between 1996 and 2004, 677 tons of fertiliziers have been used annually in Uzbekistan. More than 40% of these have been applied in the three Zarafshan provinces: Samarkand, Navoi and Bukhara. 81% of the fertilizers are N-based (127 kg/ha), 16.5% contain P_2O_5 (26 kg/ha) and 2.5% K_2O (4 kg/ha) [21]. The amount of phosphorus and potassium fertilizers has been decreasing over the recent decades, being replaced by nitrogen, which leads to an additional long-term reduction of the overall soil productivity [23].

These highly complex issues of water availability, water use efficiency, soil quality, socioeconomic growth, administrative challenges and environmental problems in the Uzbek agricultural sector have been discussed by many researchers [10] [12]-[15] [18]-[21] [24]-[31].

In the Bukhara region, however, a world heritage site and important Silk Road oasis, the water and soil resource management and the resulting environmental problems have been studied mainly by local scientists and without a broader dissemination of the results [32]-[35]. They have analyzed the local soils in great detail and worked on the problems of saline soils and the agro-physical properties and salt regime of irrigated soils. This study now focuses on the dynamics of the water resources usage and the ameliorative conditions of the irrigated areas in the Bukhara province between 2000 and 2013. The knowledge about these important aspects is essential for the improvement of the ameliorative conditions, achieving and maintaining an optimal groundwater level and controlling the groundwater salinity.

2. The Characteristics of the Investigated Region

The Bukhara region is located on the southwest of the Republic of Uzbekistan, bordered by the Kashkadarya, Navoi and Khorezm regions of Uzbekistan as well as by the autonomous republic of Karakalpakstan and the

Republic of Turkmenistan (**Figure 1**). This region covers 40,320 km² and has a population of 1.7 mln. people. The majority of the population lives in rural areas (68%) with Bukhara (0.24 mln. inhabitants) itself being the only major city in the region. The climate is continental, with cold winters (monthly average: 1.6°C in January) and hot summers (monthly average: 29.4°C in July). The annual average air temperature is 15.6°C and the annual precipitation is 142 mm. Most of the rainfall occurs during the winter months and in early spring (20.7 mm in December, 19.5 mm in January, 18.3 mm in February 28.8 mm in March) while the summer months are very dry (1.4 mm in June, 1.1 mm in July, 0.3 mm in August). Bukhara is a Silk Road oasis with a long history of cultivation. That is why only 23% of the area is unused, even though large areas of the Bukhara region are covered by the Kyzyl Kum desert, due to the arid climate with annual evaporation rates of approximately 2000 mm [11] [12]. 64% of the area is used for pastures, 4.7% for agriculture and 2.4% are covered by artificial drainage water lakes. And while the percentage of agricultural land is low, it is of great importance for the regional economy. A total of 274,900 ha are intensively irrigated to allow the production of cotton and wheat [36]. The main industries are textile, silk and cotton ginning. A second important part of the Bukhara economy is the mining and processing of oil, natural gas and precious metals. The region is one of the largest industrialized areas specializing on fuel and energy in Central Asia.

Soil Characteristics of the Bukhara Region

All of the soils found in the Bukhara region are characterized by a very low humus content (1% - 2%) [37]. The soil atlas of Uzbekistan [37] shows 11 different major soil types for this region, which can be sorted into four different categories (**Figure 2**):

- Sand and desert sand soils (**Figure 2**, No. 1 - 3; FAO-types: Dunes/Shifting Sand (DS), Cambic Arenosols (Qc) [38] [39]): These soil types dominate the western part of the Bukhara region, where the Kyzyl-Kum is located, but also the southern part, near the Uzbek-Turkmen border. These soils have a humus content of about 0.5% and nitrogen contents between 0.04% and 0.05%. Lacking both humus and nutrition elements, these soils are usually free of vegetation and thus are exposed to increased degradation and deflation.

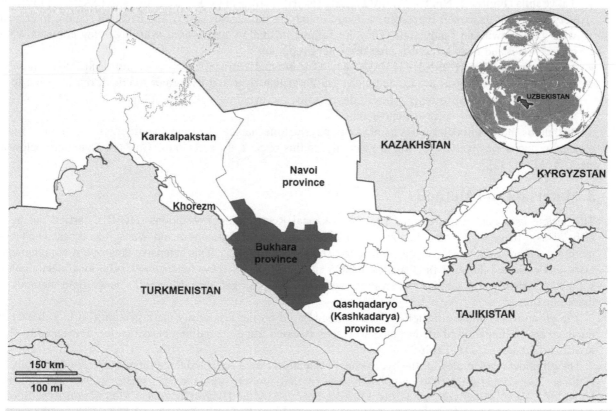

Figure 1. The map of Bukhara region (Cartography: M. Groll, Base map: commons.wikimedia.org).

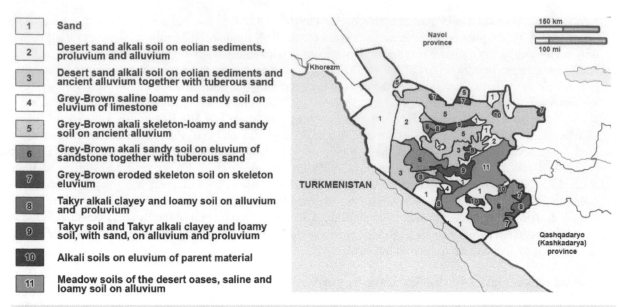

1	Sand
2	Desert sand alkali soil on eolian sediments, proluvium and alluvium
3	Desert sand alkali soil on eolian sediments and ancient alluvium together with tuberous sand
4	Grey-Brown saline loamy and sandy soil on eluvium of limestone
5	Grey-Brown akali skeleton-loamy and sandy soil on ancient alluvium
6	Grey-Brown akali sandy soil on eluvium of sandstone together with tuberous sand
7	Grey-Brown eroded skeleton soil on skeleton eluvium
8	Takyr alkali clayey and loamy soil on alluvium and proluvium
9	Takyr soil and Takyr alkali clayey and loamy soil, with sand, on alluvium and proluvium
10	Alkali soils on eluvium of parent material
11	Meadow soils of the desert oases, saline and loamy soil on alluvium

Figure 2. Soil map of the Bukhara region (Cartography: M. Groll, Base map: commons.wikimedia.org, Data: [37]).

- Grey-Brown soils (**Figure 2**, No. 4 - 7; FAO-types: Yermosols (Y)) [38]-[39]): These soil types dominate the northern and eastern parts of the region which are characterized by intensive irrigation farming on aridic soils. These soils have a varying texture (from sandy-loamy to medium-loamy) and a below average humus contents (0.6% - 0.9% in the arable layer). In the older parts of the irrigated land the amount of humus is higher and ranges from 1.2% to 1.8%. The nitrogen content in the irrigated grey-brown soils is between 0.05% and 0.16% while the total concentration of phosphor ranges from 0.09% to 0.11% [37].
- Takyr soils (**Figure 2**, No. 8 - 10; FAO-types: Takyric Yermosol (Yt), Calcic Xerosols (Xk)) [38] [39]): Not very common, these soil types can be found mainly in the central part of the Bukhara region, interlocked with grey-brown and sandy soils. They are formed in shallow depressions with high clay contents, which collect water and form salt crusts after the water evaporates.
- Meadow soils (**Figure 2**, No. 11; FAO-types: Calcaric and Eutric Gleysols (Gc, Ge)) [38] [39]): These soil types of the desert oases are located along the Zarafshan river and its former riverbed, reaching out to the Amudarya. Compared to other types of desert soils, the meadow soils have a higher humus (1.1% - 1.4%) and nitrogen (0.08% - 0.12%) content.

The humus in the soil determines its physical, physical-mechanical, hydrological, thermal, agronomic and biological properties as well as the soil productivity. In this respect, the soils in the Bukhara region are below average [14] [23] [37].

3. Materials and Methods

This article uses data from regional Hydro-Geological Reclamation Expeditions (HGRE), carried out by the Uzbek Basin Irrigation System Administration of the Ministry of Agriculture and Water Resources. The HGRE monitors the groundwater levels, mineralization and soil salinity [21]. This extensive database is complemented by results collected during 2 field studies in 2005 and 2010. During these field surveys additional data about the soil characteristics, soil salinity and cropping patterns have been gathered using soil science field methods and questionnaires.

The groundwater mineralization was determined based on the Priklonsky classification ([40], **Table 1**) and the degree of salinization of the irrigated areas was assessed using the Salinity classification developed by Bazilevich and Pankova ([41], **Table 2**).

The groundwater samples for the hydro-chemical analysis were collected three times per year (April, July and October, 2000-2013) from 1870 observation wells (metallic tubes with an inner diameter of 90 - 110 mm and a length of 3 - 6 m, filled with sand-gravel filters), operated by the Hydro-Geological Melioration Expedition of the Department of Agriculture and Water Resources in Bukhara (HGME). The samples were taken by the on-farm

Table 1. The classification of the groundwater mineralization [40].

No.	Category	Total Dissolved Solids (TDS) (g/l)
1.	Fresh	0 - 1
2.	Low mineralization	1 - 3
3.	Medium mineralization	3 - 10
4.	High mineralization	10 - 50

Table 2. The classification of the soil salinity, based on the total dissolved solids (TDS) and Clorine (Cl) [41].

No.	The Level of Salinization	Sulfate	Chloride-Sulfate		Sulfate-Chloride		Chloride
		TDS (g/l)	TDS (g/l)	Cl (g/l)	TDS (g/l)	C1 (g/l)	C1 (g/l)
1.	Non saline	<0.3	<0.1	<0.01	<0.01	<0.01	<0.01
2.	Slightly saline	0.3 - 1.0	0.1 - 0.3	0.01 - 0.05	0.1 - 0.3	0.01 - 0.04	0.01 - 0.03
3.	Moderately saline	1.0 - 2.0	0.3 - 1.0	0.05 - 0.2	0.3 - 0.6	0.04 - 0.2	0.03 - 0.1
4.	Highly saline	2.0 - 3.0	1.0 - 2.0	0.2 - 0.3	0.6 - 1.0	0.2 - 0.3	0.1 - 0.2
5.	Very highly saline	>3.0	>2.0	>0.3	>1.0	>0.3	>0.2

technicians (each responsible for a single farm with an average size of 1000 - 2000 ha), allowing the swift collection and analysis of more than 5600 groundwater samples per year. During the first sampling period (April), the effects of the extensive salt leaching, which are conducted just before the irrigation season starts, can be monitored. The second sampling period (July) covers the peak irrigation activity and the third sampling period (October) takes place immediately after the end of the growing season. This allows the analysis of the lowering of the phreatic surface without groundwater recharge. Furthermore, the analysis of the groundwater table dynamic outside the growing season is important, as a seasonal salinity restoration might occur when the upward flux prevails over the lateral outflow.

15,000 soil samples were taken by the HGME after the end of the growing season in November of each year (2000-2013) in 0 - 30 cm, 30 - 70 cm and 70 - 100 cm depth. Each sample is considered to be representative for 10 - 20 ha, resulting in a much denser grid of soil data than groundwater data. For each soil sample the electric conductivity was measured at the four corners of a 1.5×2.0 m^2 area surrounding the soil sampling site. The determination of the soil salinity was carried out applying two different methods. First through the extraction and assessment of the soluble salt contents and second through a SM-138 conductivity sensor applied to a 1:1 mixture of soil sample and water [42]. The measured soil salinity was then categorized as low (0.02 - 0.06 mg equivalent per liter), moderate (0.06 - 0.12 mg equivalent per liter) and high (>0.12 mg equivalent per liter). This classification was then assigned to the area represented by each soil sample in order to gain the spatial information about the soil salinity distribution and dynamic.

4. Results and Discussion

4.1. Development of the Irrigated Area and the Application of Fertilizers

Between 1991 and 2014, the population of the Bukhara province has grown by 47% (from 1.195 mln to 1.756 mln) [42], resulting in a strong economic growth and increasing food demand. In order to meet the growing demand the irrigated area increased as well-but at a slower rate (+12.7% since 1991; +0.4% between 2000 (273,800 ha) and 2013 (274,900 ha) while the population grew by 21.9% during the same period) [36]. This ongoing expansion of the irrigated land takes place on marginal lands with lower soil fertility and higher degrees of salinization. Thus, not only did the amount or irrigated land per capita decrease from 0.2 ha in 1991 to 0.16 ha in 2013 (−22%), but the productivity of that land also decreased, so that the crop yields in the Bukhara region (Wheat: 2.4 t/ha and year; Cotton: 2.1 t/ha and year; Vegetables and fruits: 10.6 t/ha and year) are now below the national average (Wheat: 2.5 t/ha and year; Cotton: 2.2 t/ha and year; Vegetables and fruits: 11.0 t/ha and year).

In order to increase the crop yield on the unproductive soils, mineral and organic fertilizers are applied in large quantities. **Table 3** shows a similar composition of the fertilizers in the Bukhara region and nation-wide (95% - 96% organic fertilizer), but also that on average 41.5% more fertilizers are applied in the Bukhara province (8646.7 kg/ha and year) than throughout the country (6110 kg/ha and year) and that the amount of fertilizers applied in the Bukhara region increased from 7510 kg/ha in 2011 to 9330 kg/ha in 2013 (+24.2%). And even though this increase will in the short-term stabilize the crop yields and secure the food and revenue generation of the region, in the long-term the extensive usage of (mineral) fertilizers will reduce the soil productivity and lead to more salinization [14] [21] [23].

4.2. Usage of the Water Resources

The irrigation farming is not only the most important part of the economy (90% of all crops are grown on irrigated areas) in the Bukhara region, it is also the largest water user (between 2000 and 2013 on average 95.8% of the annual water consumption has been allocated to the agricultural sector, **Figure 3**) [43]. Until the 1950s this water had been provided by the Zarafshan River, which crossed the Bukhara province and flowed into the Amu-Darya near Türkmenabat in Turkmenistan. But due to the extensive water withdrawal in the middle reaches of the Zarafshan the river since then officially "ends" 40 km northeast of Bukhara near Gijduvon, all its remaining water being distributed into the irrigation network [6]. Only during spring the Zarafshan River water is still used

Table 3. Application of mineral and organic fertilizer (in kg/ha) in the Bukhara region in comparison to the Uzbek average.

Year	Bukhara Region (kg/ha)		Uzbekistan Average (kg/ha)	
	Mineral Fertilizer	Organic Fertilizer	Mineral Fertilizer	Organic Fertilizer
2011	370	7140	260	5470
2012	380	8720	290	5850
2013	330	9000	310	6150
Average	**360**	**8286.7**	**286.7**	**5823.3**
Total	**8646.7**		**6110**	
%	**4.2**	**95.8**	**4.7**	**95.3**

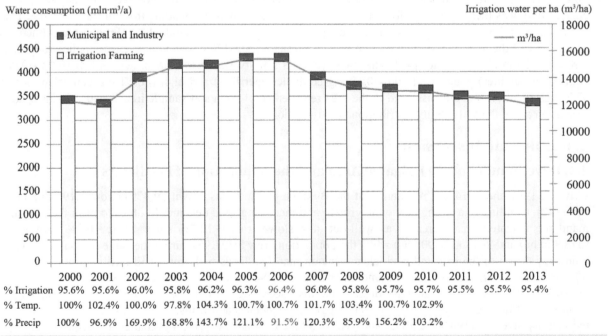

Figure 3. Water consumption in the Bukhara province between 2000 and 2013.

for the leaching of the irrigated areas. The main source of water is nowadays the Amu-Darya. Water is withdrawn and delivered to the Bukhara oasis by a series of pumping stations, overcoming a height-difference of 68 m and a distance of 194 km. The hydrographic infrastructure of the Amu-Bukhara irrigation canal network includes the II-Amu-Bukhara Machine Station, the Amu-Karakul canal as well as the Kuyi-Mazar, Tudakul and Shurkul reservoirs, which are used for the irrigation water management. This irrigation network is complemented by an extensive drainage system with six major Drainage Water Collectors (DWC)-Central, North, Parallel, Tashkuduk, Parsankul and Ogitma.

Figure 3 shows the water consumption in the Bukhara province between 2000 and 2013. The municipal and industrial water use has been consistently low throughout this period, ranging from 150.3 mln·m^3 (2001) to 177.4 mln·m^3 (2003). The irrigation water on the other hand was subject to a strong interannual dynamic (min: 3.28 km^3 in 2001, max: 4.23 km^3 in 2006, average: 3.7 km^3) (**Figure 3**). The highest water consumption in 2006 (total: 4.4 km^3) coincides with a below average precipitation (91.5% of the reference year 2000), which might have influenced the agricultural water consumption. This total high also represents the year with the highest percentage of water used for the irrigation farming (96.4%), though this percentage has been very consistent throughout the whole period (average: 95.8%). The amount of water used for irrigation per hectare or irrigated land ranged between 15,425 m^3/ha in 2005 and 11,950 m^3/ha in 2013, with an average of 13,495 m^3/ha [12] [13]. To contrast these values, in the neighboring and equally arid Navoi province 18,754 m^3/ha are consumed. In Khorezm the amount of water used for the irrigation is also higher than in the Bukhara province (15,500 m^3/ha) while in Karakalpakstan the water consumption is comparable (13,200 m^3/ha) and in the Samarkand province, which is located further upstream along the Zarafshan River and is characterized by a higher annual rainfall (339 mm) only 6504 m^3 of water are used per hectare [6] [12] [13]. But the additional rainfall is not the main reason for the much lower water consumption in the Samarkand province, but the need for extensive leaching in the more arid provinces which are more heavily affected by salinization [12] [13]. On top of that the irrigation efficiency is very low in Central Asia. Significant losses are caused by the open irrigation canals (evaporation and infiltration due to lack of maintenance), large scale reservoirs (evaporation) and by the inefficient furrow irrigation [10] [15] [21] [44]. Dukhovny and de Schutter [10] have stated that the efficiency within the Amu-Darya catchment was as low as 50% in 1965 and that in the Zarafshan River catchment this low efficiency was already undercut in 1936. A higher efficiency of up to 82% was possible in the Golodnaya Steppe through the use of modern drip irrigation methods in the 1970s, but unfortunately these modern approaches were not implemented area-wide. The performance of the irrigation and drainage networks in Central Asia further decreased considerably since the disintegration of the Soviet Union as infrastructure maintenance is no longer coordinated and the financial means are no longer available. In the Uzbek part of the Zarafshan catchment 540,000 ha of arable land are irrigated by a main irrigation system with a length of 3140 km (41% of the canal length is lined) and an interfarm irrigation system with a length of 17,400 km (11% of which are lined)- and most of this network is in dire need of rehabilitation [14].

The water consumption and meteorological data (**Figure 3**) furthermore show that in years with an above average precipitation the water consumption for the irrigation farming is also above average (100% - 125% precipitation = 116% irrigation water consumption; >160% precipitation = 118% irrigation water consumption; base year: 2000). This means that the water availability in the irrigation network (rivers, canals and reservoirs) is already today the limiting parameter for the agricultural activities in this region and that a decrease of the water availability (e.g. by climate change induced droughts) cannot be buffered by the existing reservoir capacities and will directly affect the agricultural productivity.

Besides the very high water losses because of unsuitable irrigation methods and the overall lack of maintenance of the water infrastructure, the irrigation farming is still seen as the best economic option for the arid lowlands of the Aral Sea basin. There are however, other problems related to the irrigation besides the considerable water losses. In the Bukhara region these include excessive waterlogging, soil salinization, water depletion and water quality degradation.

The drainage of the irrigated and leached fields started in 1932. Drainage water collectors (DWC) were constructed throughout the Bukhara oasis, especially between 1956 and 1979. The main collectors in the region are the West Romitan, North Bukhara, Dengizkul, Main Karakul, Parallel and Central Bukhara collectors. The drainage not only removes the surplus water from the irrigated fields but also the leached salts, fertilizers and pesticides. As a result the mineralization of collector-drainage water is very high in the Bukhara region and ranges from 3.8 g/l to 4.2 g/l. These values are way above the national threshold of 1 g/l and indicate the severe impairment of the water quality in the lower Zarafshan River catchment [6]. But the CDW quality is not uniform throughout the Bukhara province as it also reflects the groundwater and soil salinity, which show a strong spatial varibility [45].

The collected drainage water is mostly dumped into natural depressions outside of the irrigated zone. These include the Shurkul Lake (filled with water from the West Romitan, Mahankul, Gurdyush and Main Karakul collectors), the Karakyr depression (served by the Severobuhra collector) and the Agitma depression (filled by the Agitma collector).

4.3. The Groundwater Table and Its Influence on the Characteristics of the Irrigated Area

The fertility of the irrigated lands in arid regions is largely dependent on the groundwater table and its salinity. Maintaining an adequate groundwater level is therefore a critical factor in creating the best ameliorative conditions for the irrigated lands. The groundwater table (GWT) itself is mainly determined by the terrain, the drainage depth, the distance from the drains and from the infiltration and percolation of irrigation water and precipitation [24]. High water tables of saline groundwater lead to a capillary rise of the salts into the upper soil layers and to water logging in the root zone, resulting in a reduction of the crop yields. In order to secure sufficiently high yields and regain the sustainability of the agriculture a better understanding of the groundwater dynamics and the spatial distribution of salinized areas, water logging and the salinization risk. Based on such results localized measures can be taken in order to prevent any further worsening of the status quo.

The data collected from the 1870 groundwater monitoring stations shows that the majority of the irrigated areas (63.7%) in the Bukhara region have a groundwater table depth of 2.0 - 3.0 m, followed by areas with a groundwater table depth of 1.5 - 2.0 m (20.8%). On the other hand, areas with a very high water table and areas with a very low water table only made up a very small percentage of the total irrigated area (**Figure 4**). From 2000 to 2013, areas with a groundwater table depth of higher than 3 m increased their coverage from 80.1% to 94.9%. On average the groundwater table rose by 38 cm from 2.66 m (2000) to 2.28 m (2013). This means that the irrigated area with a shallow groundwater table of 3 m or less (and thus a high potential for water logging and salinization) increased from 219,300 ha in 2000 to 260,900 ha in 2013.

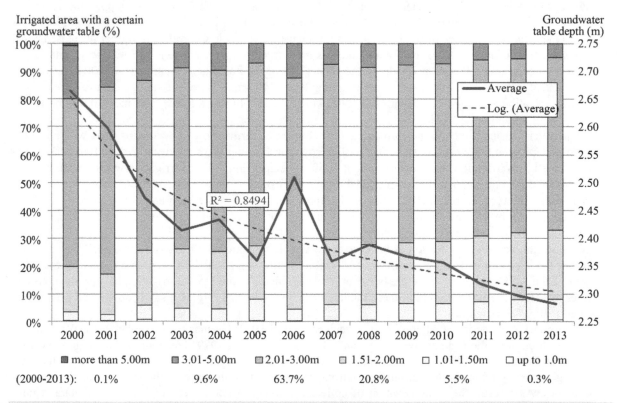

Figure 4. Temporal dynamic of the groundwater table depth in the irrigated areas (in % and average depth in m).

Figure 4 furthermore shows a drop of the average groundwater table by 15 cm in 2006, the year with the highest amount of irrigation water but a very low precipitation (comp. **Figure 3**). This indicates that the precipitation—even in this arid region—has a considerable influence on the groundwater recharge and that this influence might be greater than that of the irrigation farming. This is also supported by the fact that in the years after 2006 the amount of irrigation water decreased continuously, while the groundwater level kept rising. A third important parameter influencing the groundwater table in the Bukhara region is the inflow of groundwater from deeper aquifers, which pushes towards the surface [24].

4.4. The Groundwater Salinity and Its Impact on the Irrigated Areas

The groundwater table is only one driver of the salinization. The salinity of the groundwater is another equally important one. On average the groundwater in the Bukhara region has a salinity between 3.53 g/l (in 2001) and 2.83 g/l (in 2013) (**Figure 5**), which are extremely high values. The Amu-Darya for instance has a mineralization of 0.2 - 0.3 g/l in its middle reach and up to 1.0 g/l in the lower reach [46]. The same salinity (0.98 - 1.09 g/l) has been reported for the lower Zarafshan River near Navoi [6]. The majority of the irrigated areas in the Bukhara region showed a salinity of 1.1 - 3.0 g/l (61.1%, 168,000 ha) while the percentage of areas with a low salinity of lower than 1.0 g/l was with 0.2% (421 ha) negligible.

Between 2000 and 2013 the groundwater salinity decreased from more than 3.1 g/l to below 2.9 g/l. But just as for the water consumption for irrigation, the precipitation and the groundwater table, the groundwater salinity shows an anomaly in the year 2006. The salinity peak in the shallow groundwater was probably caused by a dry year (see **Figure 3**), where the lower amount of infiltrating water not only led to a drop of the groundwater table (see **Figure 4**), but also to higher concentrations of the applied fertilizers in the soils as the thinning effect of the rainfall has been much less pronounced than in previous years.

The contribution of the groundwater to the evaporation in the Bukhara province is between 60% and 80% and on average up to 5000 m³ of groundwater per hectare rises up into the root zone and evaporates each year [47] [48].

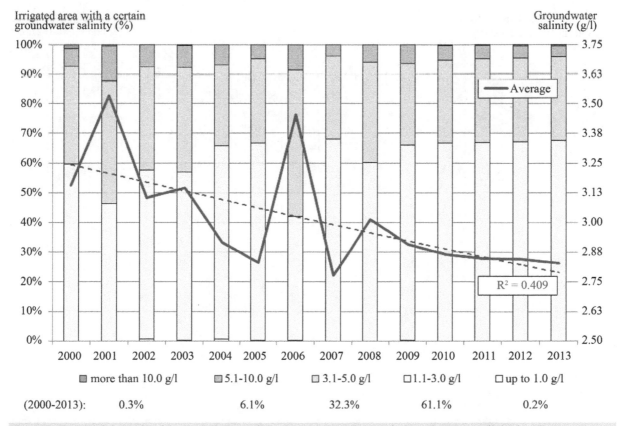

Figure 5. Temporal dynamic of the groundwater salinity in the irrigated areas (in % and average salinity in g/l).

Based on the average groundwater salinity of 3.02 g/l the annual amount of salts accumulated in the irrigated soils of the Bukhara province is 15.1 t/ha (for a total of 4.14 mln. tons per year). Broken down for the different categories of groundwater salinity this means that in areas with a groundwater salinity of 1 - 3 g/l (61.1% of the irrigated area) approximately 10 tons of salt are accumulated per hectare and year while in areas with a groundwater salinity of 5 - 10 g/l approximately 37.5 t/ha are accumulated. That means that even though the overall groundwater salinity has decreased over the past decade, certain areas within the Bukhara province are still heavily affected by groundwater based salt influx. This secondary salinization is combated by excessive leaching and increased irrigation quotas. But while these measures can help to mitigate the salt stress of the crops in the short term, in the long term this practice only leads to a further rise of the groundwater table, as only 30.2% of the irrigated areas in the Zarafshan River catchment are connected to the drainage system [6], and an increase of the groundwater salinity, especially as the irrigation water itself is characterized by a high mineralization of more than 1 g/l (the threshold for surface water mineralization in Uzbekistan) [35] [47].

4.5. The Soil Salinization in the Irrigated Areas

The soil salinization in the Bukhara province is, due to the salinity of the shallow groundwater and the irrigation water, the extensive use of fertilizers and the high evaporation rates, a widespread problem. 91.6% (251,400 ha) of all irrigated areas are at least slightly saline (**Figure 6**), 35.4% (97,200 ha) are characterized by at least a medium salinization and 7.8% (21,400 ha) are considered fully saline. These data confirm the salinization assessment by the Uzbek state committee for land resources, geodesy, cartography and state cadastre [49], which in 2008 had presented a map of the salinization in parts of the Bukhara oasis (**Figure 7**).

On average, the soil salinization score, which is calculated using the intensity of the salinization (four classes) and the percentage of the irrigated area affected by that salinization intensity and can have a score between 0 (non saline) and 1 (fully saline), is with 0.45 on a moderate level. The temporal development of this score shows a peak in 2001 (0.524), followed by a steady reduction of the soil salinization until 2009 (0.417). Since then the salinization again slightly increased. This trend does not correlate statistically with the average air temperature

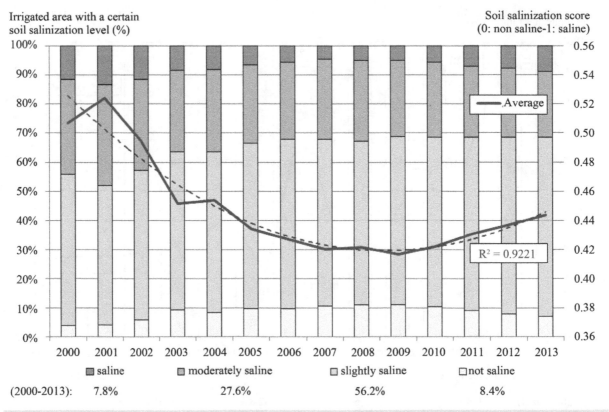

Figure 6. Temporal dynamic of the soil salinization in the irrigated areas (in % and soil salinization score between 0 (non saline) and 1 (saline)).

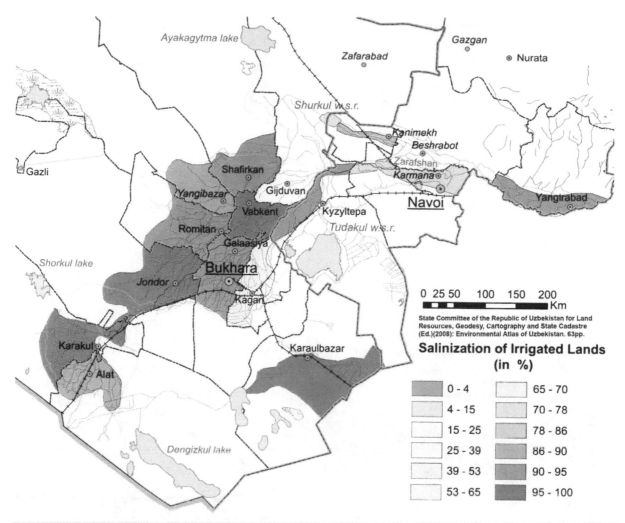

Figure 7. Soil salinization in the Bukhara oasis [49].

or the precipitation (R^2_{temp} = 0.028; R^2_{precip} = 0.059; **Figure 8**), though the strong decrease of the soil salinization in 2002 and 2003 coincides with above average rainfall (359 mm and 353 mm) in the same period. A better connection has been identified between the soil salinization and the groundwater salinity (R^2_{gwsal} = 0.38; **Figure 8**) in that the soil salinization for several years (2001-2004) followed the development of the groundwater salinity with a one-year delay (see **Figure 5** & **Figure 6**). But the best correlation can be seen between the soil salinization and the groundwater table (R^2_{gwtab} = 0.65). An especially good match is the rising groundwater level and the increase of the soil salinization since 2009 (see **Figure 4** & **Figure 6**). The irrigation intensity (in m³/ha) on the other hand did not show a clear connection to the soil salinization (R^2_{irrig} = 0.41). The data indicate that the soil salinization slightly decreases with an increase of the irrigation water consumption, but this probably only symbolizes the short-term effect of the leaching and does not reflect the long-term soil deterioration caused by extensive irrigation in arid regions.

The data presented here indicates that several factors are influencing the soil salinity and that the weight of those factors might shift from year to year. Depending on the overall factor constellation, the rainfall (2002-2003), the groundwater salinity (2001-2004) or the groundwater table (2009-2013) determine the further development of the soil salinization. Based on this complex system, excessive leaching is not an effective tool for the stabilization of the irrigated areas, especially as the irrigation water consumption does not correlate as well with the soil salinity as the other parameters (with exception of the average air temperature). In order to prevent a further spread of the soil salinization in the Bukhara region the water logging and the groundwater table have to be managed in more efficient ways. Potential measures for this could be:

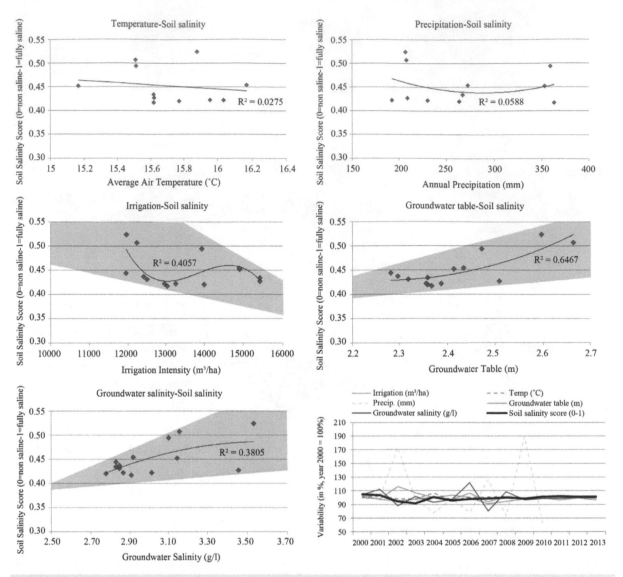

Figure 8. Correlations between the soil salinization and its influencing parameters.

- Increasing the density of the drainage systems so that a larger percentage of the irrigated area can be drained;
- Better maintenance and rehabilitation of the hydraulic structures in order to increase the efficiency of both the irrigation and the drainage system;
- Increasing the water quality of the irrigation water (in the lower Zarafshan River catchment untreated drainage water has to be used for irrigation purposes as there is an annual water deficit of 1.6 km^3) [6];
- A more sustainable use of the water and land resources (irrigation and fertilization based on the demand rather than on a quota).

4.6. Soil Conditions of Reclaimed Irrigated Lands

As the agricultural sector is of great importance for the Uzbek economy, protecting and improving the condition of the irrigated areas is of national interest, as can be seen in the presidential decree 3932 "on measures to radically improve the system of land reclamation" [50]. The actions taken for the rehabilitation of the irrigated areas result in an overall mostly "satisfactory" condition of the reclaimed land (74.3% of all reclaimed areas). A smaller portion (7.7%) even has a "good condition" while the reclaimed land with an "unsatisfactory condition" cover 18.1% of the area (**Figure 9**). Between 2000 and 2008 the average condition of the reclaimed areas improved slightly (from a score of 0.434 to 0.47), but since then the condition was deteriorating again (back to 0.426).

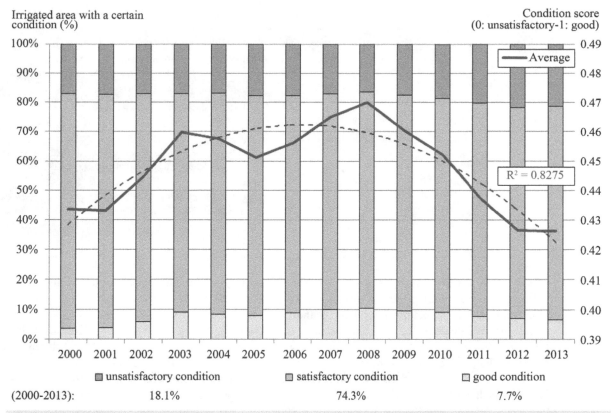

Figure 9. Temporal dynamic of the condition of the reclaimed irrigated areas (in % and condition score between 0 (unsatisfactory) and 1 (good)).

The temporal development of the condition of the reclaimed land shows a good correlation to the irrigation intensity (in m³/ha, R^2_{irrig} = 0.7, see **Figure 3**). The recent worsening could furthermore be explained by the raising groundwater table and the accelerated spread of the soil salinization since 2009 (see **Figure 4** and **Figure 6**). Especially the shallow groundwater is critical in certain areas, as the crop yields show significant losses if the groundwater table is at 1.5 m or less [47]. This threshold was in 2013 exceeded by 8.2% of the irrigated area (22,500 ha), while in 2000 only 3.25% of the irrigated area (8900 ha) had been affected by this.

The crop yield losses are aggravated by the soil salinity, as can be seen in **Table 4**. Both wheat and cotton— the two dominant crop types in the Bukhara province are not well suited for the cultivation on saline soils, as they suffer crop losses of 50% or more even on moderately saline soils, which make up 27.6% of all irrigated areas in the region. On soils with an even higher salinity (7.8% of all irrigated areas) the yield losses are higher than 80%. For cotton this translates to yields of 0.9 - 1.25 t/ha on saline soils in comparison to 1.6 - 2.1 t/ha on non-saline soils. Other crop types, like beets, corn, alfalfa or eggplants can tolerate higher salt concentrations. Especially beets can be grown on highly saline soils and still produce yields of more than two thirds of their maximum capacity [51].

5. Conclusions

Due to the arid climate the agriculture in the Bukhara region depends heavily on extensive irrigation. But as the flat topography does not provide sufficient natural drainage, water logging and raising groundwater tables are serious problems for the agricultural productivity. The combination of the high salinity of the irrigation water and the generous application of fertilizers leads to a widespread soil salinization. Excessive leaching is supposed to reduce the top soil salinity, but as the drainage system is only covering a small portion of the irrigated areas and is in need of maintenance, this process only contributes to the ongoing salinization and the reduction of soil fertility and crop yields.

The data shown here for the years 2000 to 2013 indicate that the groundwater table is rising throughout the

Table 4. The influence of the soil salinity on crop yields (modified after: [51]).

Type of crops	Crop yields (%) for different soil salinity classes				
	Non-Saline	Slightly Saline	Moderately Saline	Highly Saline	Very Highly Saline
Cotton	100	94	50		
Wheat	100	80	39		
Corn (fodder)	100	98	72	57	35
Alfalfa	100	96	73	53	39
Potato	100	90	68		
Tomato	100	98	74	54	34
Pea	100	66	27		
Eggplant	100	92	74	48	32
Beet	100	95	88	73	66

region while the groundwater salinity is decreasing. The soil salinity on the other hand is, after an improvement during the first half of the study period, slightly increasing since 2009, which is also reflected in the slight worsening of the condition of the reclaimed land during the same period.

The groundwater-soil-meteorology-irrigation system is highly complex, so that not a single parameter is controlling the salinization process at all times. Different parameters seem to dominate the system at different times, which also means that such a simple solution as excessive leaching will not work in the long term.

In order to manage the groundwater table and the soil salinity more effectively, the data from the existing monitoring network have to be implemented into the agricultural practices, so that the water usage can be tailored to the actual demand and the on-site capacity. Further advisable actions include the rehabilitation and extension of the drainage system, an increase of the irrigation efficiency, the improvement of the irrigation water quality and the consideration of more salt-tolerant crop types.

Acknowledgements

Prof. Rashid Kulmatov is grateful for the German Academic Exchange Service (DAAD), Germany, for funding the scientific exchange, under which the present research was carried out.

References

[1] Pankov, E.I., Aydarov, I.P., Yamnova, I.A., Novikova, A.F. and Blagovolin, N.S. (1996) Natural Zonation of Saline Soils in the Aral Sea Basin (Geography, Genesis, Evolution). Nauka, Moscow. (In Russian)

[2] Panin, P.S. (1968) The Salt-Efficiency Processes in Leached Soil Strata. Nauka, Novosibirsk. (In Russian)

[3] Agaltseva, N. (2008) Assessment of the River Extreme Runoff for Climate Scenarios Conditions. NIGMI Bulletin # 7, Climate Change Effects, Adaptation Issues, Tashkent, 5-9.

[4] Agaltseva, N. and Chub, V. (2010) Methodical Approach to the Assessment of Climate Change Impact on Water Resources in Uzbekistan. In: Merz, B., Dukhovny, V. and Unger-Shayesteh, K., Eds., *Water in Central Asia*, *Volume of Abstracts*, *International Scientific Symposium*, Tashkent, 24-26 November 2010, 9.

[5] Chub, V., Agaltseva, N. and Myagkov, S. (2002) Climate Change Impact on the Rivers Runoff for the Central Asian River. Hydrometeoizdat, Tashkent.

[6] Groll, M., Opp, C., Kulmatov, R., Ikramova, M. and Normatov, I. (2015) Water Quality, Potential Conflicts and Solutions—An Upstream-Downstream Analysis of the Transnational Zarafshan River (Tajikistan, Uzbekistan). *Environmental Earth Sciences*, **73**, 743-763. http://dx.doi.org/10.1007/s12665-013-2988-5

[7] Ibatullin, S., Yasinsky, V. and Mironenkov, A. (2009) Impacts of Climate Change on Water Resources in Central Asia. Sector report of the Eurasian Development Bank, Almaty.

[8] Lioubimtseva, E. and Henebry, G.M. (2009) Climate and Environmental Change in Arid Central Asia: Impacts, Vulnerability, and Adaptations. *Journal of Arid Environments*, **73**, 963-977. http://dx.doi.org/10.1016/j.jaridenv.2009.04.022

[9] Spektorman, T. and Petrova, E. (2008) Application of the Climatic Indices for the Assessment of Climate Change Im-

pact on the Health of Uzbekistan Population. NIGMI Bulletin # 7, Climate Change Effects, Adaptation Issues, Tashkent, 37-46.

[10] Dukhovny, V. and de Schutter, J.L. (2011) Water in Central Asia–Past, Present, Future. CRC Press, New York.

[11] Chub, V. and Ososkova T. (2008) Second National Report of the Republic of Uzbekistan on UN FCCC. Tashkent.

[12] Alihanov, B.B. (2008) About a Condition of Environment and Use of Natural Resources in Republic of Uzbekistan (The Retrospective Analysis for 1988-2007). National Report of the State Committee for Nature Protection of the Republic of Uzbekistan, Tashkent. (In Russian)

[13] Umarov, N.U. (2013) About a Condition of Environment and Use of Natural Resources in Republic of Uzbekistan (The Retrospective Analysis for 2008-2011). National Report of the State Committee for Nature Protection of the Republic of Uzbekistan, Tashkent. (In Russian)

[14] Bucknall, J., Klytchnikova, I., Lampietti, J., Lundell, M., Scatasta, M. and Thurman, M. (2003) Irrigation in Central Asia—Social, Economic and Environmental Considerations. World Bank Report.

[15] Abdullaev, I., Kazbekov, J., Manthritilake, H. and Jumaboev, K. (2009) Participatory Water Management at the Main Canal—A Case from South Ferghana Canal in Uzbekistan. *Agricultural Water Management*, **96**, 317-329. http://dx.doi.org/10.1016/j.agwat.2008.08.013

[16] Aparin, V., Kawabata, Y., Ko, S., Shiraishi, K., Nagai, M., Yamamoto, M. and Katayama, Y. (2006) Evaluation of Geoecological Status and Anthropogenic Impact on the Central Kyzylkum Desert (Uzbekistan). *Journal of Arid Land Studies*, **15**, 129-133.

[17] Crosa, G., Stefani, F., Bianchi, C. and Fumagalli, A. (2006) Water Security in Uzbekistan—Implication of Return Waters on the Amu Darya Water Quality. *Environmental Sciences and Pollution Research*, **13**, 37-42. http://dx.doi.org/10.1065/espr2006.01.007

[18] Fayzieva, D.K., Atabekov, N.S., Usmanov, I.A., Azizov, A.A., Steblyanko, S.N., Aslov, S. and Khamzina, A.S. (2004) Hydrosphere and Health of Population in the Aral Sea Basin. In: Fayzieva, D.K., Ed., *Environmental Health in Central Asia—The Present and Future*, WITPRESS, Southampton, 51-100.

[19] Fedorov, Y.A., Kulmatov, R.A. and Rubinova, F. (1998) The Amudarya, in a Water Quality Assessment the Former Soviet Union. In: Kimstach, V., Maybeck, M. and Baroudy, E., Eds., *A Water Quality Assessment the Former Soviet Union*, E&FN Spon, London and New York, 413-445.

[20] Toderich, K.N., Abbdusamatov, M. and Tsukatani, T. (2004) Water Resources Assessment, Irrigation and Agricultural Developments in Tajikistan. Kyoto Institute of Economic Research, Discussion Paper No. 585, Kyoto.

[21] MAWR (2008) Annual Reports 2004-2008. Ministry of Agriculture and Water Resources, Tashkent. (In Russian)

[22] World Bank (2010) Uzbekistan—Climate Change and Agriculture—Country Note. Washington.

[23] FAO (2003) Fertilizer Use by Crop in Uzbekistan. Land and Plant Nutrition Management Service & Land and Water Development Division, Rome.

[24] Kulmatov, R. (2014) Problems of Sustainable Use and Management of Water and Land Resources in Uzbekistan. *Journal of Water Resource and Protection*, **6**, 35-42. http://dx.doi.org/10.4236/jwarp.2014.61006

[25] Awan, U.K., Ibrakhimov, M., Tischbein, B., Kamalov, P., Martius, C. and Lamers, J.P.A. (2011) Improving Irrigation Water Operation in the Lower Reaches of the Amu Darya River—current Status and Suggestions. *Irrigation and Drainage*, **60**, 600-612. http://dx.doi.org/10.1002/ird.612

[26] Kulmatov, R.A. (1994) Laws of Distribution and Migration of Toxic Elements in River Waters of the Aral Sea Basin. Tashkent. (In Russian)

[27] Kulmatov, R.A. and Hoshimhodjaev, M. (1992) Spatial Distribution and Speciation of Microelements in Water of Zarafshon River. *Water Resource*, **11**, 103-114. (In Russian)

[28] Kulmatov, R.A. and Nasrulin A.B. (2006) Spatial-Temporary Distribution of Polluting Substances in Amu-Darya River. *Water Problems of Arid Territories*, **6**, 20-30. (In Russian)

[29] Saito, L., Fayzieva, D., Rosen, M.R., Nishonov, B., Lamers, J. and Chandra, S. (2010) Using Stable Isotopes, Passive Organic Samplers and Modeling to Assess Environmental Security in Khorezm, Uzbekistan. Final Report, NATO Science for Peace Project No. 982159.

[30] Scott, J., Rosen, M.R., Saito, L. and Decker, D.L. (2011) The Influence of Irrigation Water on the Hydrology and Lake Water Budgets of Two Small Arid-Climate Lakes in Khorezm, Uzbekistan. *Journal of Hydrology*, **410**, 114-125. http://dx.doi.org/10.1016/j.jhydrol.2011.09.028

[31] Shanafield, M., Rosen, M., Saito, L., Chandra, S., Lamers, J. and Nishonov, B. (2010) Identification of Nitrogen Sources to Four Lakes in the Agricultural Region of Khorezm, Uzbekistan. *Biogeochemistry*, **101**, 357-368. http://dx.doi.org/10.1007/s10533-010-9509-3

[32] Lebedev, Y.P. (1954) The Soils of Irrigation Oases in the Lower Zarafshan River. Aral-Caspian Expedition, Moscow, 1. (In Russian)

[33] Zvonkova, T.V. (1965) The Bukhara Region—The Environmental Conditions and Resources of the South-Western Uzbekistan. *Science*, **3**, 333-346. (In Russian)

[34] Kimberg. N.V. (1975) The Soil Classification—Soils in Uzbekistan, Tashkent.

[35] Abdullaev, S. (1975) The Agro Physical Properties and Salt Regime of Irrigated Soils of the Bukhara Region. PhD Thesis, Institute of Soil, Tashkent. (In Russian)

[36] State Department of Statistics of Uzbekistan (2014) Annual Statistics 2000-2014. (In Russian) www.statistics.uz

[37] State Committee of the Republic of Uzbekistan for Land Resources, Geodesy, Cartography and State Cadastre (2012) The Soil Atlas of Uzbekistan. Tashkent. (In Russian)

[38] FAO (1974) Soil Map of The World—1:5,000,000, Volumes 1-10. Food and Agriculture Organization of the United Nations and UNESCO, Paris.

[39] FAO (1988) Soils Map of the World—Revised Legend. Food and Agriculture Organization of the United Nations and UNESCO, Rome.

[40] Priklonskiy, V. (1970) The Methodical Recommendations on Mineralization of Solonetz Lands and Accounting Saline Soils. Nauka, Moscow. (In Russian)

[41] Bazilevich, N.I and Pankova, E.I. (1970) Classification of Groundwater According to the Degree of Mineralization. Moscow. (In Russian)

[42] State Committee of the Republic of Uzbekistan on Statistics (2015) Demographic Information—Number of Resident Population 1991-2014. www.old.stat.uz/en/demographic/

[43] Kulmatov, R. (2008) The Modern Problems in Using, Protecting and Managing Water and Land Resources of the Aral Sea Basin. In: Qi, J. and Evered K.T., Eds., *Environmental Problems of Central Asia and Their Economic, Social and Security Impacts*, Springer Netherlands, Dordrecht, 24-32. http://dx.doi.org/10.1007/978-1-4020-8960-2_2

[44] Olsson, O., Ikramova, M., Bauer, M. and Froebrich, J. (2010) Applicability of Adapted Reservoir Operation for Water Stress Mitigation under Dry Year Conditions. *Water Resources Management*, **24**, 277-297. http://dx.doi.org/10.1007/s11269-009-9446-x

[45] Shodiev, S.R. (2009) The Hydrochemistry Rivers and Drainage Water Southwest of Uzbekistan. PhD Thesis, Institute of Hydrometeorology, Tashkent. (In Russian)

[46] Kulmatov, R. and Hojamberdiev, M. (2010) Speciation Analyses of Heavy Metals in the Transboundary Rivers of Aral Sea Basin—Amudarya and Syrdarya Rivers. *Journal of Environmental Science and Engineering*, **4**, 36-45.

[47] Eshchanov, R. (2008) The Agro Ecological Bases for Sustainable Use of Land and Water Resources (On Example of Khorezm Region). PhD Thesis, Thesis, Institute of Soil, Tashkent. (In Russian)

[48] Shirokov, Y.I. and Chernyshev, A.K. (1999) Rapid Method for Determination of Soil Salinity and Water in Condition of Uzbekistan. *Journal of Agriculture of Uzbekistan*, **5**, 45-52. (In Russian)

[49] State Committee of the Republic of Uzbekistan for Land Resources, Geodesy, Cartography and State Cadastre (2008) Environmental Atlas of Uzbekistan, Tashkent.

[50] President of the Republic of Uzbekistan (2007) Decree UP-3932 from October 29th on Measures to Radically Improve the System of Land Reclamation. (In Russian)

[51] SANIIRI (2005) Improving of Monitoring of Salt Processes on Irrigated Lands through the Use of Modern Technologies and the Development of Ways to Prevent Crop Damage from Soil Salinity. Final Report, Tashkent. (In Russian)

11

Fate of Nutrients, Trace Metals, Bacteria, and Pesticides in Nursery Recycled Water

Yun-Ya Yang, Gurpal S. Toor*

Soil and Water Quality Laboratory, Gulf Coast Research and Education Center, Institute of Food and Agricultural Sciences, University of Florida, Wimauma, FL, USA
Email: *gstoor@ufl.edu

Abstract

Faced with rapid population growth and fresh water scarcity, reuse of reclaimed water is growing worldwide and becoming an integral part of water resource management. Our objective was to determine the fate of nutrients, trace metals, bacteria, and legacy organic compounds (organochlorine pesticides) in the recycled water from five commercial nursery ponds in Florida. The pH of recycled water at all sites was 8.1 - 9.3, except one site (6.5), while the electrical conductivity (EC) was 0.31 - 0.36 dS/m. Concentrations of trace metals in recycled water were low: Fe (0.125 - 0.367 mg/L), Al (0.126 - 0.169 mg/L), B (0.104 - 0.153 mg/L), Zn (0.123 - 0.211 mg/L), and Mn (<0.111 mg/L). Total phosphorus (P) and total nitrogen (N) in the recycled water were 0.35 - 1.00 mg/L and 1.56 - 2.30 mg/L, respectively. Among organochlorine pesticides, endrin aldehyde was the only pesticide detected in all nursery recycled water ponds, with concentrations from 0.04 to 0.10 μg/L at four sites and 1.62 μg/L at one site. Other detected pesticides in recycled water were methoxychlor, endosulfan sulfate, dichlorodiphenyldichloroethylene (DDE) and α-chlorodane, with concentrations < 0.20 μg/L. Total coliforms and *Escherichia coli* (*E. coli*) in recycled water were 20 - 50 colony forming units (CFU)/100 mL. We conclude that the concentrations of various inorganic and organic compounds in recycled water are very low and do not appear to be problematic for irrigation purposes in Florida's nursery recycled water ponds.

Keywords

Water Quality, Recycled Water, Nutrients, Trace Metals, Pathogen, Pesticides

1. Introduction

In the world, 60% to 90% of available water is used for agricultural purposes [1]. In the United States, approx-

---

*Corresponding author.

imately 80% of the consumptive water is used in agriculturally related activities [2]. Growing population is increasingly competing with commercial and agricultural uses for limited freshwater supplies. In future, agricultural and other industries may have to heavily rely on using recycled water to meet crop irrigation needs. Water reuse is a growing practice worldwide and Florida leads the United States in using reclaimed or recycled water [3]. Reclaimed water is the domestic wastewater that has received at least secondary treatment and basic disinfection. In Florida, reclaimed water has been safely and successfully used since 1966 [4]. In 1992, Florida produced 1.1 billion liters of reclaimed water per day and this doubled to more than 2.7 billion liters per day by 2013 [5]. The practice of using reclaimed water for irrigation purposes can result in savings of freshwater supplies and can partially supply some nutrients, particularly nitrogen (N) and phosphorus (P). However, there are concerns about the impact of the quality of the reclaimed water on crops, soils, and irrigation systems.

Recycling ponds are often regarded as a best management practice for eliminating potential problems that arise from container nursery runoff [6]. Chemicals runoff from nursery production facilities is a concern because it is a potential nonpoint source of pollution. Runoff of residential chemicals may be considerable in some situations [7]-[9]. For example, a six-state survey of container nurseries found that nitrate-N (NO_3-N) levels in runoff, irrigation ponds and wells exceeded the US Environmental Protection Agency's (US EPA) drinking water limit of 10 mg/L [10].

Due to the continuous reuse, water in the nursery recycling ponds can contain high concentrations of nutrients and soluble salts and may also contain trace metals, pathogens, and pesticides. The presence of excess amounts of these constituents in recycled water may have adverse effects on plant growth and environment, if runoff water reaches water bodies of concern [8] [10]. Therefore, it is important to know if the recycled water contains harmful compounds in excessive amounts, and if so, what mitigation approaches are needed to remove these compounds from the recycled water, so that it can be continuously used in an environmentally sustainable way. In addition, monitoring of physical, chemical, and biological variables in recycled water is necessary to ensure acceptable water quality. The results from a survey in greenhouse and nursery irrigation water highlighted that monitoring of physical, chemical, and biological water quality can be used to improve irrigation system design when implementing a water treatment technology [11].

As in other parts of the world, nursery and landscape industry in Florida is currently facing the challenge of managing an economically important sector in an environmentally-sustainable way. It is important to be proactive to find if there are any concerns related to the presence of chemical compounds in the recycled water and if so develop approaches to alleviate those concerns. Documenting the safe use of recycled water will be of significant benefit to not only assure the sustainability of agricultural practices but also assure future environmental compliance. Thus, the overall goal of this study was to determine the fate of major water quality constituents in nursery recycled water. The specific objective was to characterize salinity, nutrients, trace metals, bacteria, and pesticides in recycled water samples from five commercial nurseries in Florida that represented diversity in plant species production and management practices.

2. Materials and Methods

2.1. Site Description

Five commercial (working) nurseries with a recycled pond were used for this study. The recycled ponds in these nurseries receive any water leaving the property (nursery site) following a significant rainfall and/irrigation event. All nurseries were located in the Hillsborough County, which drains into the Tampa Bay Watershed in the west coast of Florida. Due to the confidentiality agreement signed with the nursery growers, we cannot disclose the names and specific locations of nurseries. The county has an area of 2714 km^2. Of that area more than 84% (2350 km^2) is unincorporated. The Hillsborough County land use/land cover consists of about 46% urban and built up, 19% agriculture and 17% wetlands. The remaining 18% is designated as recreational, open land, rangelands, upland forests, water and other uses [12].

2.2. Sample Collection

Water samples were collected from five commercial nursery ponds in April 2010 from 8 a.m. to 12 p.m. Three of the ponds were located within one nursery and two ponds were located at two other nurseries. The climate in the area is subtropical, with an average annual air temperature of 22°C and daily extremes ranging from 0.7°C to 35°C and total rainfall varies from 120 to 130 cm per year.

Water samples were collected using a long-handled polyethylene 500 mL capacity dipper (SCIENCEWARE®). Samples for organochlorine pesticide analysis were collected in 1-L amber glass bottles (Fisher Scientific, Fair Lawn, NJ). Samples for nutrients, trace metals, and bacteria were collected in separate 250 mL high-density polyethylene (HDPE) bottles (Fisher Scientific, Fair Lawn, NJ). All samples were stored at 4°C until analysis. At each pond, samples were also analyzed in-situ for pH, electrical conductivity (EC), temperature, and dissolved oxygen (DO) using a Manta Water Quality Multiprobe (Eureka Environmental Engineering, Austin, TX).

2.3. Sample Preparation and Analysis

2.3.1. Nutrients, Cations, and Trace Metals

Approximately 100 mL of each collected water sample was preserved with concentrated H_2SO_4 or HNO_3 at pH 2 for total P and trace metals analysis, respectively. Samples filtered through 0.45-μm filter paper were analyzed for chloride (Cl^-) and ammonium-N (NH_4-N) and nitrate/nitrite-N (NO_x-N; $NO_2^- + NO_3^-$) using an Automated Discrete Analyzer (AQ2+, Seal Analytical Inc., Mequon, WI) with EPA method 325.2 [13] and 350.1 [14], respectively. Total N was determined in unfiltered water samples using alkaline persulfate digestion method [15] and then analyzed for NO_x-N as above. Organic N was calculated as the difference between total N and (NH_4-N + NO_x-N). Preserved water samples were analyzed for total P, cations, and trace metals including aluminum (Al), boron (B), calcium (Ca), copper (Cu), iron (Fe), potassium (K), magnesium (Mg), manganese (Mn), molybdenum (Mo), sodium (Na), and zinc (Zn) using an inductively coupled plasma-optical emission spectrometer (ICP-OES; PerkinElmer Optima 2100 DV; PerkinElmer, Shelton, CT) according to EPA method 6010 C [16].

2.3.2. Pesticides

Organochlorine pesticides analysis was conducted using EPA Method 3510 C [17]. In brief, 1 L of sample was extracted with 60 mL of methylene chloride (Fisher Scientific, Fair Lawn, NJ) three times in 2-L separatory funnel. Sample bottles were extracted rinsed with methylene chloride to recover any adsorbed analyte. The separatory funnel was sealed and shaken vigorously for 2 min and allowed to settle. After complete separation, the organic phase was drained by centrifugation at 3000 rpm for 5 min (Sorvall Legend RT, Thermo Scientific), while the aqueous phase was re-extracted. The extracted organic phase was combined and concentrated to 1 mL using a Caliper Life Sciences Turbo Vap II Concentration Evaporator (Turbo Vap II, Zymark Inc.). Samples were then transferred onto a Resprep Florisil cartridge (3 mL, 250 mg), and eluted with 50 mL of hexane (Fisher Scientific, Fair Lawn, NJ). The eluent was collected and evaporated to 5 mL and analyzed with a gas chromatography—electron capture detector (GC-ECD) system.

A PerkinElmer Clarus 500 GC together with a Restek Rtx-CL Pesticides column (30 m × 0.32 mm id × 0.5 μm; Restek Corp, USA) was used for the separation and analysis of the following organochlorine pesticides: aldrin, α-hexachlorocyclohexane (HCH), β-HCH, γ-HCH, δ-HCH, α-chlordane, γ-chlordane, 4,4'-dichloro-di-phenyl-dichloroethane (DDD), 4,4'-dichloro-diphenyl-dichloroethylene (DDE), 4,4'-dichloro-diphenyl-trichloroethane (DDT), dieldrin, endosulfan I, endosulfan II, endosulfan sulfate, endrin, endrin aldehyde, endrin ketone, heptachlor, heptachlor epoxide, and methoxychlor. The surrogates used were 2, 4, 5, 6-tetrachloro-mxylene (TCMX) and decachlorobiphenyl (DCB). The temperature of injector and detector were 250°C and 425°C, respectively. The column temperature program was as follows: initial temperature 100°C held for 2 min, ramped to 200°C at 45°C min^{-1}, and held for 1 min. The oven temperature was then ramped to 300°C at 15°C min^{-1} and held for 3 min. The column pressure was held at 25 psi (helium) and the sample was injected in a splitless mode. Confirmation of the chlorinated pesticides was done by running the same samples on a Rtx-CL Pesticides 2 column (30 m × 0.25 mm id × 0.2 μm; Restek Corp., USA) under slightly different conditions. The slight adjustment to the temperature program was made to optimize peak splitting between δ-HCH and heptachlor. The changes that were made were to the final oven temperature ramp from 300°C at 15°C min^{-1}, held for 3 min to 325°C at 8.5°C min^{-1} and held for 2 min. Another change was that the column pressure was held at 11 psi throughout the run.

2.3.3. Total Coliforms and *Escherichia coli*

Total coliforms and *Escherichia coli* (*E. coli*) concentrations in pond waters were quantified via membrane filtration on MI agar BBLTM (Becton Dickinson) using EPA method 1604 [18].

3. Results and Discussion

3.1. Basic Properties of Recycled Water

The pH is an important characteristic of water when considering reuse of recycled water for irrigation. It is well documented that the ideal pH for most plant production is 5.0 - 6.5 as this provides a balance of availability of essential plant nutrients [19]. The pH of recycled water at four sites was alkaline (8.1 - 9.3) and one site had neutral pH (6.5) (**Table 1**). In general, the measured values of pH were within the EPA recommended range (6 - 9) of water reuse for irrigation [20]. The pH values from sites C and D were slightly higher than the upper limit of the recommended range.

Electrical conductivity (EC) is directly related to the concentration of dissolved ions in the water. The EC values at all sites were 0.311 - 0.362 dS/m, which was equivalent to 199 - 232 mg/L of total dissolved solids (TDS; **Table 1**). The EC and TDS levels were well below the EPA limits of <2.0 dS/m EC and 500 - 2000 mg/L TDS in the irrigation water for nursery crops [20]. The water temperature in the ponds fluctuated from 21.4°C to 25.3°C at the sampling time. Other chemical properties of the recycled water included 2 - 12 mg/L DO, 23% - 145% saturated DO, and 9.2 - 25.8 mg/L Cl^- (**Table 1**); these values are in fact better than the typical values found in most stream waters, suggesting that the recycled water is of good quality.

3.2. Concentrations of Cations in Recycled Water

Total cations concentrations in the recycled water ranged from 62.7 to 84.3 mg/L; of which, Ca^{2+} was the dominant cation accounting for 52% - 73% (39.3 - 55.3 mg/L) (**Table 2**). The concentration of other cations, *i.e.*, Mg^{2+}, K^+, and Na^+ were 5 - 11.5, 4.4 - 16.7, and 6.2 - 18.5 mg/L, respectively.

3.3. Concentrations of Trace Metals in the Recycled Water

Concentrations of Fe, Al, Zn, B, and Mn were less than 0.4 mg/L in all pond waters (**Table 3**). Copper (Cu) and

Table 1. Selected physical and chemical properties (mean ± standard deviation) of nursery recycled water (n = 4).

Site	pH	EC (dS/m)	TDS[a] (mg/L)	Cl^- (mg/L)	DO (mg/L)	Saturated DO (%)	Temperature (°C)
A	8.9 ± 0.12	0.33 ± 0.0024	211 ± 1.5	11.3	7.8 ± 1.08	95 ± 13.9	25.3 ± 0.19
B	8.1 ± 0.19	0.311 ± 0.008	199 ± 0.5	13.2	7.1 ± 0.19	86 ± 2.4	25.3 ± 0.12
C	9.3 ± 0.04	0.328 ± 0.0012	210 ± 0.7	12.6	7.2 ± 0.27	87 ± 4.1	25.2 ± 0.12
D	9.1 ± 0.22	0.362 ± 0.0085	232 ± 5.4	25.8	12.0 ± 1.6	145 ± 20	24.5 ± 0.71
E	6.5 ± 0.08	0.317 ± 0.001	203 ± 0.6	9.2	2.0 ± 0.25	23 ± 2.7	21.4 ± 0.23

[a]Total dissolved salts (EC (dS/m) × 640).

Table 2. Concentrations of cations in nursery recycled water[a].

Site	Ca^{2+} (mg/L)	Mg^{2+} (mg/L)	K^+ (mg/L)	Na^+ (mg/L)	Total cations (mg/L)
A	40.1 (56)	11.5 (16)	9.1 (13)	10.7 (15)	71.3
B	39.6 (59)	9.1 (13)	8.3 (12)	10.6 (16)	67.6
C	39.3 (52)	8.8 (12)	16.7 (22)	11.4 (15)	76.2
D	55.3 (66)	5.0 (6)	5.5 (7)	18.5 (22)	84.3
E	45.9 (73)	6.3 (10)	4.4 (7)	6.2 (10)	62.7

[a]Values in parentheses are percent of total cations.

Table 3. Concentrations of metals in nursery recycled water.

Site	Fe (mg/L)	Al (mg/L)	Zn (mg/L)	B (mg/L)	Mn (mg/L)
A	–[a]	0.169	0.149	–	–
B	0.180	0.177	0.211	0.153	–
C	–	0.126	0.123	0.138	–
D	0.125	0.149	0.154	0.104	–
E	0.367	–	0.162	–	0.111

[a]Not detected.

Mo were not detected in any of the samples. Values of detected trace metals were below EPA's recommended range of reclaimed water quality for irrigation (*i.e.*, 5 mg/L of Al and Fe; 0.2 mg/L of Cu and Mn; 2 mg/L of Zn) [20].

3.4. Concentrations of Nitrogen and Phosphorus Forms in the Recycled Water

Total N concentrations in the recycled water were 1.56 - 2.30 mg/L; of which, 78% to 98% was organic N (1.42 - 1.88 mg/L) with the remainder (2% - 22%) being in inorganic N forms such as NO_3-N + NO_2-N (not detected, ND-0.35 mg/L), and NH_4-N (0.03 - 0.37 mg/L; **Table 4**). Total Kjeldahl N (TKN) ranged from 1.47 to 2.17 mg/L. The NH_4-N concentration at site E was 10 times greater than other sites likely due to the low DO level as compared to other sites (see **Table 4**). The less aerobic conditions (*i.e.*, low DO) present at the site E may have reduced the oxidation of NH_4^+ to NO_3^-, resulting in higher NH_4^+ concentrations. Across all sites, concentrations of NH_4-N were inversely correlated ($R = -0.86$) with DO.

Total P concentrations were 0.59 - 1.00 mg/L in the recycled water, of which, 87% - 96% was present as soluble inorganic P (PO_4-P, 0.32 - 0.94 mg/L) with the remainder (4% - 13%) as organic P forms (**Table 4**).

3.5. Concentrations of Pesticides in Recycled Water

Five legacy organochlorine pesticides were detected in the recycled water, with concentrations below 2 µg/L (**Table 5**). Among the analyzed pesticides, endrin aldehyde (0.044 - 1.619 µg/L) was detected at all sites. Methoxychlor (ND-0.193 µg/L) and endosulfan sulfate (ND-0.157 µg/L) were found at three and two sites, respectively. Alpha-chlordane (ND-0.046 µg/L) and DDE (ND-0.053 µg/L) were only found at one site. These organochlorine pesticides found in pond waters may have been due to the past usage as most of them were banned in the United States in the 1970s. The past usage of these pesticides likely resulted in presence of these in the pond waters with eroded soil particles.

3.6. Total Coliform and *Escherichia coli* in the Recycled Water

Total coliform and *E. coli* concentrations in recycled water were 22 - 47 and 20 - 35 CFU/100 mL, respectively (**Figure 1**). These values of total coliform and *E. coli* were within the safe range recommended by EPA (<1000 CFU/100 mL) [20]. Low values of total coliforms in recycled water, which can be used as an indicator of total load of pathogens, suggests a negligible risk of spreading of plant diseases with reclaimed water irrigation.

Table 4. Concentrations of nitrogen and phosphorus forms in nursery recycled water.

Site	NO_2-N + NO_3-N (mg/L)	NH_4-N (mg/L)	Organic N (mg/L)	TKN (mg/L)	Total N (mg/L)	PO_4-P (mg/L)	Total P (mg/L)
A	0.35 (15)[a]	0.04 (2)[a]	1.88 (83)[a]	1.92	2.26	0.51 (87)[b]	0.59
B	0.10 (7)	0.05 (3)	1.42 (90)	1.47	1.57	0.32 (92)	0.35
C	0.07 (3)	0.07 (3)	1.85 (93)	1.92	1.98	0.94 (94)	1.00
D	–[c]	0.03 (2)	1.53 (98)	1.56	1.56	0.48 (89)	0.54
E	0.13 (6)	0.37 (16)	1.80 (78)	2.17	2.30	0.70 (96)	0.73

[a]Values in parentheses are percent of total N; [b]Values in parentheses are percent of total P; [c]Not detected.

Table 5. Concentrations of five detected organochlorine pesticides in nursery recycled water.

Site	Endrin aldehyde (µg/L)	Methoxychlor (µg/L)	Endosulfan sulfate (µg/L)	DDE (µg/L)	α-Chlordane (µg/L)
A	1.619	0.193	0.014	0.053	–[a]
B	0.078	–	0.157	–	0.046
C	0.044	0.078	–	–	–
D	0.082	–	–	–	–
E	0.102	0.035	–	–	–

[a]Not detected.

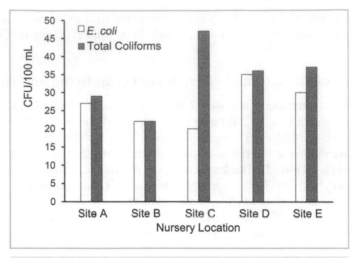

Figure 1. Total coliforms and *E. coli* in nursery recycled water.

4. Conclusion

Characterization of recycled water from commercial nurseries can aid in understanding if there are any contaminants of concern in recycled water for plant productivity and for environmental pollution if excess runoff leaves the nursery sites. Our data show that recycled water from nursery ponds is of good quality with low levels of nutrients, trace metals, bacteria, and legacy pesticides. Low levels of nutrients and trace metals in recycled water can still be beneficial for the plants, as many of these are essential elements for plant growth and productivity. Organic N and soluble inorganic P are the dominant forms of nutrients in recycled water. The presence of high organic N in nursery ponds shows that N is efficiently cycled in the ponds. These results indicate that recycled water in nursery ponds is not polluted and the risk of spreading of plant diseases in nurseries using reclaimed water is negligible.

Acknowledgements

We are thankful to Florida Nursery Growers and Landscape Association (FNGLA) for providing funding for this research work and nursery growers for allowing access to the recycled water ponds.

References

[1]   Oron, G., Gillerman, L., Buriakovsky, N., Bick, A., Gargir, M., Dolan, Y., Manor, Y., Katz, L. and Hagin, J. (2008) Membrane Technology for Advanced Wastewater Reclamation for Sustainable Agriculture Production. *Desalination*, **218**, 170-180. http://dx.doi.org/10.1016/j.desal.2006.09.033

[2]   US Department of Agriculture (2013) Economic Research Service: Irrigation & Water Use. http://www.ers.usda.gov/topics/farm-practices-management/irrigation-water-use.aspx#.U--cdPldV4g

[3]   Parsons, L.R., Sheikh, B., Holden, R. and York, D.W. (2010) Reclaimed Water as an Alternative Water Source for Crop Irrigation. *HortScience*, **45**, 1626-1629.

[4]   Southwest Florida Water Management District (SFWMD) (2009) Reclaimed Water: A Reliable, Safe Alternative Water Supply. https://www.watereuse.org/files/s/docs/reclaimed_water_lev2_08_09.pdf

[5]   Florida Department of Environmental Protection (FDEP) (2014) 2013 Reuse Inventory. Florida Department of Environmental Protection, Tallahassee, Florida. http://www.dep.state.fl.us/water/reuse/docs/inventory/2013_reuse-report.pdf

[6]   Fain, G.B., Gilliam, C.H., Tilt, K.M., Oliver, J.W. and Wallace, B. (2000) Survey of Best Management Practices in Container Production Nurseries. *Journal of Environmental Horticulture*, **18**, 142-144.

[7]   Ristvey, A.G., Lea-Cox, J.D. and Ross, D.S. (2004) Nutrient Uptake, Partitioning and Leaching Losses from Container-Nursery Production Systems. *Acta Horticulturae*, **630**, 321-328.

[8]   Mangiafico, S.S., Gan, J., Wu, L.S., Lu, J.H., Newman, J.P., Faber, B., Merhaut, D.J. and Evans, R. (2008) Detention and Recycling Basins for Managing Nutrient and Pesticide Runoff from Nurseries. *HortScience*, **43**, 393-398.

[9]   Colangelo, D.J. and Brand, M.H. (2001) Nitrate Leaching Beneath a Containerized Nursery Crop Receiving Trickle or Overhead Irrigation. *Journal of Environmental Quality*, **30**, 1564-1574. http://dx.doi.org/10.2134/jeq2001.3051564x

[10] Yeager, T., Wright, R., Fare, D., Gilliam, C., Johnson, J., Bilderback, T. and Zondag, R. (1993) Six State Survey of Container Nursery Nitrate Nitrogen Runoff. *Journal of Environmental Horticulture*, **11**, 206-208.

[11] Meador, D.P., Fisher, P.R., Harmon, P.F., Peres, N.A., Teplitski, M. and Guy, C.L. (2012) Survey of Physical, Chemical, and Microbial Water Quality in Greenhouse and Nursery Irrigation Water. *HortTechnology*, **22**, 778-786.

[12] Hillsborough Community Atlas (2014)
      http://www.hillsborough.communityatlas.usf.edu/general/default.asp?ID=12057&level=cnty.

[13] US Environmental Protection Agency (1983) Method 325.2. Chloride by Automated Colorimetry. Methods for the Chemical Analysis of Water and Wastes (MCAWW).
      http://www.caslab.com/EPA-Methods/PDF/EPA-Method-3252.pdf

[14] US Environmental Protection Agency (1993) Method 350.1. Determination of Ammonia Nitrogen by Semi-Automated Colorimetry. Environmental Monitoring Systems Laboratory, Office of Research and Development, USEPA, Cincinnati. http://water.epa.gov/scitech/methods/cwa/bioindicators/upload/2007_07_10_methods_method_350_1.pdf

[15] Ebina, J., Tsutsui, T. and Shirai, T. (1983) Simultaneous Determination of Total Nitrogen and Total Phosphorus in Water Using Peroxodisulfate Oxidation. *Water Research*, **17**, 1721-1726.
      http://dx.doi.org/10.1016/0043-1354(83)90192-6

[16] US Environmental Protection Agency (2007) Method 6010 C (SW-846). Inductively Coupled Plasma-Atomic Emission Spectrometry, Revision 3. http://www.epa.gov/osw/hazard/testmethods/sw846/pdfs/6010c.pdf

[17] US Environmental Protection Agency (1996) Method 3510C. Separatory Funnel Liquid-Liquid Extraction, Revision 3. http://www.epa.gov/osw/hazard/testmethods/sw846/pdfs/3510c.pdf

[18] US Environmental Protection Agency (2002) Method 1604. Total Coliforms and *Escherichia coli* in Water by Membrane Filtration Using a Simultaneous Detection Technique (MI Medium). Rept. Mo. EPA 821-R-02-024, Environmental Protection Agency, Washington DC. http://www.epa.gov/microbes/documents/1604sp02.pdf

[19] Ghehsareh, A.M. and Samadi, N. (2012) Effect of Soil Acidification on Growth Indices and Microelements Uptake by Greenhouse Cucumber. *African Journal of Agricultural Research*, **7**, 1659-1665.

[20] US Environmental Protection Agency (2004) Guideline for Water Reuse. EPA 645-R-04-108, Environmental Protection Agency, Washington DC. http://water.epa.gov/aboutow/owm/upload/Water-Reuse-Guidelines-625r04108.pdf

12

Environmental Anthropological Study of Watershed Management-Water Quality Conservation of Forest as a Catchment Area in the Southern Part of Australia

Akira Hiratsuka[1*], Yugo Tomonaga[2], Yoshiro Yasuda[3]

[1]Osaka Sangyo University, Osaka, Japan
[2]National Museum of Ethnology, Osaka, Japan
[3]University of Hyogo, Kobe, Japan
Email: [*]hiratuka@ce.osaka-sandai.ac.jp

Abstract

Authors have conducted an experiment of irradiation using sound waves (frequency) including ultrasonic waves into water such as drinking water, sea water and forest water and wastewater so far. As a result, almost the same effect of improvement of water quality was confirmed for each sound wave. Then, an environmental anthropological study of watershed management based on the sound was carried out assuming that a water quality management using the sound could be possible. The Goulburn River basin in the southern part of Australia in which indigenous peoples (Yorta Yorta) have been concerned with the management for a long time so far was selected as an objective drainage basin this time. As a result, a couple of environmental anthropological perspectives on watershed management were proposed.

Keywords

Watershed Management, Water Quality Conservation, Water Quality Improvement by Sound, Murray-Goulburn River, Eco-Money (Environmental Tax)

1. Introduction

At present, the importance of sustainable water quality conservation of watershed/ forest as a catchment area has

---
[*]Corresponding author.

been pointed out all over the world [1] [2]. It is thought that if sustainable water quality conservation of watershed/forest is possible, the supply of drinking water with an abundant quantity, an inexpensive price and a good quality has also become a sustainable one [3]. For this purpose, a construction of "environment-social system" by the independent effort of water quality conservation driven by watershed community residents is required. It is necessary for that to make the system involved in the community for conserving the watershed. For instance, a construction of value-led management and a planning of environment investment strategy which become a mortgage of sustainability are required. Then, we carried out the survey of water quality conservation by the sounds at the indigenous communities nearby Murray-Goulburn river watershed in the southern part of Australia. And, an environmental anthropological consideration on a concrete and ideal method (watershed culture) on a construction of "environment-social system" by the independent effort of water quality conservation driven by the watershed community residents was done. Furthermore, based on the consideration, the proposal toward a water quality conservation of watershed/forest as a catchment area has been conducted, and also has been examined on the approach required in contrast with the independent effort of water quality conservation driven by the watershed community residents.

2. History of Natural Resource Management in the Study Area [4]

The main areas of our study are the Goulburn Valley that contains following towns around 200 km up to the north from Melbourne city; Echca, Moama, Barmah, Cummeragunga Aboriginal community, and Mooroopna and Shepparton along the Golburn River diverging from Murray River. The Barmah forest is a part of the largest River Red Gum wetland in the world and the size of the forest including the Millawa forest on the New South Wales side is around 30,000 hectors. The forest has more than 385 species of indigenous flora and 273 species of indigenous fauna and is listed as a wetland of international significance under the Ramsar Convention. The Barmah forest contains many sites of cultural significance to the Aboriginal people. The study areas including the Barmah forest abound in the primal industry such as fruits, vegetables, rice, dairies, livestock and sawmills. Most of people running the forest and agricultural businesses are the descendants of the Anglo-Celt settlers who had immigrated into this area since 1840s (**Figure 1**) [5].

The primal industry, specifically the forest industry slashing and processing trees for rail way and for paddle steamer yielded the first output of all production in 1860s. In 1870s, Victoria colonial government enforced the Drought Proof Policy, which controls flood in Murray-Goulburn Broken Catchment. Thus, many regulators and water ways were constructed for irrigation and agricultural and farm industry increased output as second output of all production in this area. However excessive cropping by irrigation led to the exhaustion of the soil and the big cattle and sheep station got rabbit damage and extreme reduction of the market price. After enforcing the water policy for irrigation rights among three states; New South Wales, Victoria and South Australia, more facilities such weirs and dams for water allocation were constructed but these facilities caused to drought and saltation for agriculture. Consequently, disastrous drought has attacked the native fauna and for a high salinity was found since 1930s [6]. Three state governments did not coped with these issues until 1980s because they treated resources in the forest and river in the Murray-Goulbern Valley not as a part of eco system but as a part of productive resources [7]. In 1972, United Nations Conference on the Human Environment was held in Stockholm and environmental issues become international agenda thus, natural conservation was the

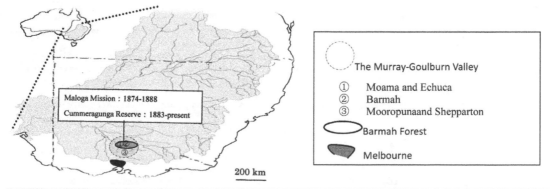

Figure 1. Study area: Murray-Goulburn Valley.

controversial issue among the Australian citizens and academics and then, state governments in Australia become enforced many measures for environmental protection. Under the circumstances, some practical measures such cooperative and joint management with the indigenous people for the environmental protection in the Murray-Goulbern Valley has been set [8]. Therefore, indigenous peoples in the area have a political voice for the negotiation on the forest and river management with local, state and federal governments without any denial at present.

3. State of Water Quality for the Goulburn Broken Catchment (1995-2010) [9]

Figure 2 shows a Goulburn Broken Catchment area. The area of the Catchment is 24,310 km^2. And, **Figure 3** shows a Goulburn River basin. Red numbers show the river reaches. The water sampling points in this time are shown in this figure. Furthermore, **Figure 4** shows the water quality monitoring sites by the authority in the Goulburn River region. According to the Goulburn River Water Monitoring Report 1995-2010, a summary regarding the water quality (Turbidity, Electrical Conductivity (Salinity), Dissolved Oxygen, pH and Phosphorus) for the Goulburn Broken Catchment is as follows [10]. **Table 1** shows the summary of the state of water quality (1995-2010) for the Goulburn Broken Catchment briefly.

Therefore, it can be said that efforts need to be made by the people centering on the community residents to avoid the deterioration of Total Phosphorus at present.

Figure 2. Goulburn broken catchment [10].

Table 1. State of water quality for the Goulburn Broken Catchment (summary).

Water quality items measured	State of water quality
1) Turbidity	The value of Turbidity is temporarily high by the influence of fire at the upper reach. But, it progresses toward recover after 2007.
2) Electrical conductivity	The value of Electrical conductivity is in fairly in good excluding two or three sites of the creek.
3) Dissolved oxygen	The value of Dissolved oxygen is in fairly in good at the all sites.
4) pH	The value of pH is in stable in the basin.
5) Total phosphorus	The value of Total Phosphorus shows the deterioration at almost all sites since the end of 2009 and almost all sites are rated as "Poor" or "Degraded" in 2010.

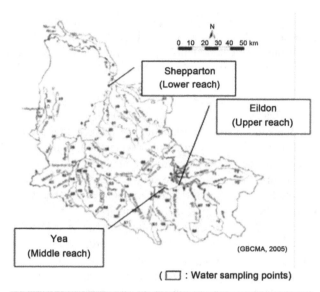

Figure 3. River reaches in the Goulburn River Basin [10].

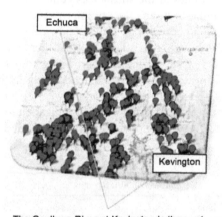

The Goulburn River at Kevington is the most
upstream site, and Echuca is the most downstream.
(Red mark: Water quality monitoring sites)

Figure 4. Water quality monitoring sites by the authority in the
Goulburn River region [10].

4. The Approach for the Water Quality Conservation by Visualization of Water Quality—A Proposal of Water Quality Mapping at Water Resource Area by Sound Waves

Authors have carried out the experiment on the improvement of water and waste-water treatment process using various sound waves (frequency) including ultrasonic wave [11] [12]. Specifically, various sound waves such as 1) ultrasonic wave; 2) music box and 3) wind-bell were irradiated to the six kinds of the water such as drinking water (hard water/soft water), forest water (upper reach/middle reach/lower reach) and sea water for removing contaminants such as nitrate, phosphorus and BOD/COD. The conclusion of the paper is as follows [12].

　1) Various sound waves affect on the improvement of the water quality.

　2) As for the removal rate of NO_2^- and NO_3^- concerning to the three samples such as the sea water, drinking water and the forest water, high values (>50%) were obtained, respectively.

　3) PO_4^{3-} was also confirmed to decrease more than 50% in the forest water.

　4) BOD and COD were drastically decreased especially in drinking water and forest water.

　5) The removal rate of Ca^{2+} and Mg^{2+} was also large (>50%) in the drinking water. Along with these, TH also drastically decreased. On the other hand, Na^+, K^+ and NH_4^+ showed very few change.

6) Therefore, decrease of NO_2^-, NO_3^- and PO_4^{3-} are due to the reaction with bivalent cations of Ca^{2+} and Mg^{2+} rather than monovalent cations such as Na^+, K^+ and NH_4^+.

7) It is supposed that if hard water with components of Ca^{2+} and Mg^{2+} are properly added into a soft drinking water, and the sound waves are irradiated to them, NO_2^-, NO_3^- and PO_4^{3-} could be reduced. It could be useful from the industrial viewpoint.

Furthermore, from the results mentioned above, a possibility of the improvement of water quality by murmuring sound of a natural stream was suggested [13] (**Figure 5**). If such a phenomenon may occurs, it is confirmed that natural purification mechanism could work. This is a subject for the near future.

Then, an environmental anthropological consideration of watershed management based on the sounds (frequency) was undertaken considering whether water quality management due to the sounds can be conducted or not in this study. This time, Goulburn river basin in the southern part of Australia was selected as an objective catchment area. One of the selected reasons is that the management of this basin has been carried out by the indigenous Australians (Yorta Yorta people) for quite a long time so far. The water sampling points in the catchment area are shown in **Figure 3**. And, the ultrasonic waves that are the easiest way to operate among the sounds were selected as an irradiated sound (frequency). The result of irradiated ultrasonic waves into Goulburn river water is shown in **Table 2**. As is seen in **Table 2** and **Figure 6**, a result of the water quality improvement of Goulburn River shows almost the same tendency as the case of Inukai River in Japan (12). Regarding nitrogen (N) and phosphorus (P), as the concentration of water quality at the sampling points (upper reach, middle reach and lower reach) was not detected, the values of the substances could not be obtained this time. It is estimated that if there exist some concentration in the river, an effect of improvement will be expected. As is seen in **Table 1** (The state of water quality for the Goulburn Broken Catchment), the most important thing is that just only phosphorus (P) in the river reduces. Therefore, it is suggested that there will be a possibility of expecting the water quality improvement using sound waves including murmuring sound in the Goulburn River. **Figure 7** shows an image picture on mapping of water quality by residents living in the watershed. If the residents will practice the water quality mapping, as a result, a conservation of water resources for drinking water will be expected through the recognition of the three factors, that is, 1) visualization → 2) awareness → 3) circulation of connection.

Table 2. Experimental results of Goulburn River water sample.

Substances	Goulburn River		
	Eildon	Yea	Shepparton
	(Upper Reach)	(Middle Reach)	(Lower Reach)
pH	7.5 → 7.0	7.5 → 7.0	7.5 → 7.5
TH (mg/L)	10 → 0	10 → 0	20 → 35
COD (mg/L)	8 → 3	8 → 3	13 → 10
BOD (mg/L)	4 → 3	4 → 3	5 → 2
NH_4^+ (mg/L)	0.26 → 0.26	0.26 → 0.26	0.26 → 0.26
NO_2^- (mg/L)	0.033 → 0.017	0.033 → 0.017	N.D.
NO_3^- (mg/L)	N.D.	N.D.	N.D.
PO_4^{3-} (mg/L)	N.D.	N.D.	N.D.
Ca^{2+} (mg/L)	12.5 → 8.5	12.5 → 8.5	15.6 → 8.8
Mg^{2+} (mg/L)	4.1 → 8.2	4.1 → 8.2	14.3 → 14.3
Na^+ (mg/L)	4.0 → 4.4	6.7 → 6.6	12.0 → 11.0
K^+ (mg/L)	0.9 → 0.9	1.4 → 1.5	2.1 → 2.3

N.D.: not detected.

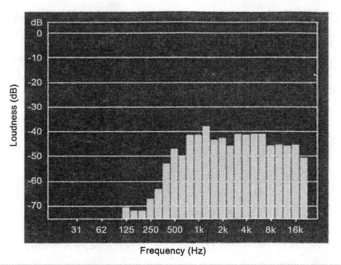

Figure 5. Example of frequency analysis of murmuring sound of stream [13].

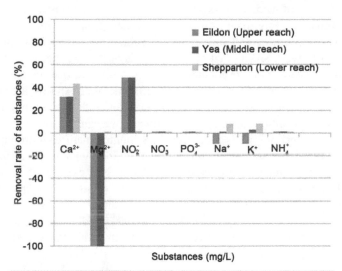

Figure 6. Removal rate of substances.

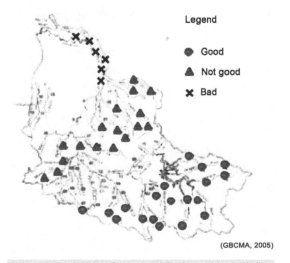

(GBCMA, 2005)

Figure 7. Mapping of water quality by residents living
in the watershed (image picture).

5. Thought of FUBEN-EKI (Further Benefit of a Kind of Inconvenience) [14] Environment-Social Systems

In this chapter, we think how we do the content of Chapter four toward the sustainable thing for the resident living in the basin. Here, a thought of "FUBEN-EKI Environment-Social Systems" which is very important in considering "sustainability" is described.

5.1. Bi-Directional Connection between Forest and Humankind Linked through Community Currency (Environmental Tax)—Construction of Values-Driven Management

First of all, it is claimed that the sending out of the following statement, that is, "Environmental conservation is incompatible with an economic growth and we should not make environmental conservation compatible with an economic growth [15] [16]" is the most important thing for solving environmental issues. Because the economic growth is based on the rent, and on the other hand, the environmental conservation is due to the ethics, respectively. That is to say, the basic principle of the former is the independence and that of the latter is the autonomy, essentially. Then which should we prioritize, the environmental conservation or the economic growth? If we think the economics is a frame of governing a nation and providing relief to people, the rent based on the money that one pay regularly for the use of land, a house or a building should be prioritized. In Japan however, there is a voice of anxiety for the situation "Environmental studies is prosperous; yet environment is ruined" [17]. So if economics is prosperous; yet country (nature/environment) is ruined, we have got our priorities wrong. In Japan, on the other hand, there is a proverb saying that "the country is destroyed; yet mountains and rivers remain." [18] We interpret this proverb as follows; the country (social system) is ruined, which means the artificial creations are surely extinct; yet the recoverable environment is still remaining. And we are sure that our country will be prosperous. As we mentioned above, environmental conservation and an economic growth are incompatible with each other. But even if the country (social system) is ruined, we can recover the environment by conservation activity again. However, if "economics is prosperous yet country (nature/environment) is ruined", then we can't recover our country. Environmental conservation (Owner: "God" which indicates the existence of an ultimate mental entity behind the universe = Owner does not exist. Not existed) is an obligation (responsibility).This is defined in the basic environment law in Japan. How can we utilize the environment? Then a concept of a rental fee to use the environment comes out. As for this, in general the existence of the user (people, company) and the owner (administration; country, city etc.) can be considered temporarily. However, a possession of the management (approval of management costs) is necessary to maintain the situation of utilization sustainably/forever.

5.2. Introduction of Community Currency (Environmental Tax) and Its Use—Planning of Environment Investment Strategy for Which Becomes a Mortgage of Sustainability

In the context of a possession of the management mentioned above, a thought of cost-benefit comes out. We use both of them as the following meanings. That is, the cost here has a meaning of doing the management expressly, that is, "sacrifice" which shows the meaning of "Kosten" in German language, and the benefit has a meaning of a favor we can use freely forever. Therefore, we think that there is a responsibility as the obligation for conserving the environment. For that, a thought of the necessity of betting money (investment) comes out. As for the investment, there are three features as follows ; 1) To be very good at appearance (to have a correspondence ability); 2) To be able to realize a return; and 3) To be able to use the eco-money. As for return ($\neq$profit, margin), useful good (good 1, good 2, good 3) as a social return comes out. A medley of those goods is environmental goods. This is called wealth. The wealth becomes full. Then, such a situation could lead to an increase of affluence (a scale of measuring the providing of relief to people). This "affluence" is just saying the worth of a nation (environmental affluence as a good), does not give the time value of currency not at all. Therefore, the currency should be used without interest. Furthermore, as for the use of eco-money (without interest) as money to spend, although this one is accompanied by an inconvenience (FUBEN-EKI) [14], it is effective for environmental conservation. There are three reasons why we use the eco-money: a) Constant & Long term (Eco-money is kept constant sustainably); b) Unreasonable (Do not be seized with common sense) and c) Non-boom (Do not boom) (see **Figure 7**). From the above, the important thing in the thought of FUBEN-EKI [14] environment-social systems is fundamentally based on the composition of a strife between "rent" and "ethics." As for rent, there are diverse rents as shown in the word "different" (a compound word between differ and rent). But strife should persist in abiding by the ethics. At present, there have been loudly called for SR (Social Responsibility), but this means a strife to abide by the ethics. From expressing in a line of connection of this sentence, SR accrues to us at the situation to keep (utilize) FUBEN-EKI [14] in an unchanged condition. We think that this is a foundation of a thought of environment-social systems in watershed management.

6. Orderly Use of Forest and Water Resource Management

6.1. Recent Trend of the Amount of Domestic Water in Use in Japan

Figure 8 shows the recent trend (1991-2010) of the amount of domestic water in use in Japan [15]. It can be seen from this figure that the amount of domestic water in use keeps on decreasing since its peak in 1997 in Japan. And, it is said that the citizens who drink a tap water as a drinking water are on a declining trend of 30 percent of the whole in Japan (Tokyo) [15]. **Figure 9** shows the amount of production and import for the PET-bottle in Japan. It can be seen from this figure that the PET-bottle as a drinking water is on an increasing trend year by year. It is thought that this decreasing trend of the water consumption has a possibility of a correlation between the citizens who do not use it as a drinking water and the trend. In addition to that, it is said according to the prediction of the long term trend of the total population in Japan that the decrease of the total population will become conspicuous since its peak in 2004 [19]. When looking at these trends, there is still some doubt about making the purified water (drinking water) abundantly using the rapid filtration in the future. When considering about the decrease of the population and the maintenance (including the countermeasure of chlorination) of the water pipe and so on, the continuous use of the present purification system (rapid filtration process) in the future is bring about the possibility of causing the high cost. On the other hand, the maintenance of the slow sand filtration becomes easy, and cost is relatively cheap to use the treatment at minimal disinfection cost. And, considering that the quantity of water does not matter appreciably with the decrease of the population, the slowness of the filtration speed does not matter. So it is thought that the effect of the purification by the creatures will be expected for that. In general, the slow sand filtration can widely remove pollutants in comparison with the rapid sand filtration. And it is thought that the both sand filtrations are realistically difficult to remove the Nitrate-N from the point of view of cost effectiveness in the future. Therefore, it is thought that the possibility of removing the fundamental pollutants by only slow sand filtration will be high if we could carry out the conservation of the water resource area sufficiently and sustainably.

6.2. Slow Sand Filtration

Slow sand filtration (ecological purification system) has been developed in the UK around two hundred years ago (see **Figure 10**). We get the drinking water with safe, cheap and delicious natural water using this method.

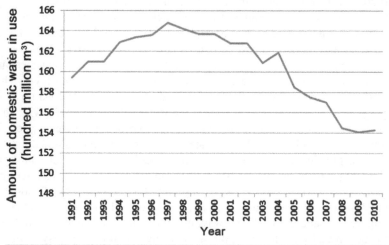

Figure 8. Changes in amount of domestic water in use (1991-2010) [15].

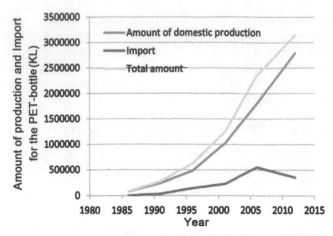

Figure 9. Changes in amount of production and import for the PET-bottle in Japan (1986-2012) [16].

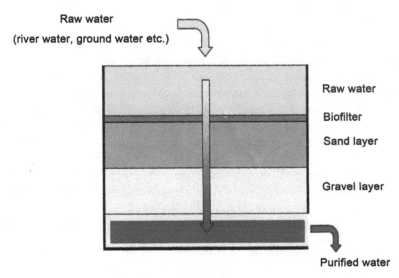

Figure 10. Schematic diagram of water purification mechanism by slow sand filtration [12].

The point of this method is just only through the raw water in the sand layer slowly. This is based on the idea (hint) that natural springs at the foot of mountains (forests) water and the riverbed are very pure. This system has once thought that the raw water is purified by just only physical filtration. The fact that the raw water is purified by the microbial action etc. has clarified around 100 hundred years ago. **Figure 11** shows a schematic diagram of water purification mechanism by slow sand filtration [12]. When passing the raw water into the slow sand filtration bed, a bio-filter by all microorganisms etc. is naturally formed on the sand layer. The raw water is purified by the microbial action etc. in the filtration bed. It is also described by N. Nakamoto as follows. "Slow sand filtration is not simply a physical filtration process. Drinking water can be produced with an ecological purification system. Biological activity in slow sand filtration is the key to producing safe and delicious drinking water. Biological activity is the basis for safe and delicious water. Algae, small animals and microscopic organisms are all important for ecological purification systems. It is emphasized that this is not simply a physical filtration process and is an ecological purification system. And, it is also stated that biological activity is the key to produce good drinking water. That is, organisms are very important for ecological purification systems. In this system, even bacteria can be removed by biological activity. On the other hand, there is no removing in rapid sand filters because of the big size of sand grain. By the way, it is often said that Japan is a mountainous country. The 70% of land of Japan is occupied by the mountainous area. Therefore, it is very easy to find spring water. Some springs have delicious water which is not only no treatment but also no relation to the chlorination. Natural

Figure 11. Eco-money to be used for conserving the water resource (image picture).

springs at the foot of mountains are delicious because of the water purified by natural processes. There are selected good water quality regions called "MEISUI 100 SEN" which was selected by the Environment Agency in Japan (1985). The criteria for selection (two essential conditions) are as follows. 1) Existence of good condition from the view point such as water quality, water quantity, surrounding environment, affinity for water. 2) Existence of conservation activity by community resident. But, the water resource area has gradually deteriorated in the region which is not sustainably active in the conservation of water environment. As a result, the region which cannot be called a MEISUI (good area for water quality) from the view point of "drinking water" is gradually increased [20]. Therefore, a construction of the wholesome environmental social system on the conservation of the water quality of water resource area is urgently required. On the other hand, rapid sand filtration plant consists of rapid sand filters, sedimentation tanks and chemical mixing basin in general. Although the area of rapid sand filters is small, a large area is necessary to sludge handling. Therefore, this system is fast indeed, but requires the treatment of large amount of the sludge as an industrial waste which is discharged from the plant every day. The cost of the treatment is getting higher than that of the water treating. That is, sludge disposal is expensive and troublesome. All microorganisms etc. cannot live in a rapid filter system and chemical reagents such as coagulants in the rapid sand filters are fatal to the creatures. It is said that chlorination was enforced by American military in Japan. They say that chlorinated tap water is not safe. However, it is said in London that non chlorinated tap water is the best water to drink. In general, the phrase "Chlorinated tap water is toxic to fish" is now common knowledge in the world. A problem on cryptosporidium flowed from the cattle pastures around a reservoir which is the water source occurred in rapid sand filters at an earlier time. But in the case of slow sand filters, cryptosporidium was not dangerous at all.

6.3. Role of the Forest as Slow Sand and Alternative Biological Filtration Equipment in Natural World

As mentioned above, this slow sand filtration holds good traditional technology even today in the world. This technology is old, but also thought wise use of natural steps for human life. Thorough investigation on the relationship between murmuring sounds and water quality in the mountains (forests), we have to learn the importance of the role of the forest in natural world. Colin Walker, who is an elder of the Yorta Yorta Aboriginal group, told the meaning of river and forest. "Our lakes, we got Barmah lake and Moira lake. When you look at the map, it looks like the living things because lakes look like kidneys and river running down through the lakes in the middle were the spine. If you look at properly you will see it looks like human body. Little creeks through running into the forest look like veins. Veins on the water get on the running in purifying the lakes just our vein, our body, and our blood. That is purified by two lakes as the kidneys" (Walker, C. Interview with Y. Tomonaga, his house at Cummeragunga, March 24 2008) (see **Figure 12**). In connection with this, there is an unique book entitled "Toward Water Reuse: Learn It from Your Own Body (1976)" written by Dr. Isu. Kyu [21]. In the case of the Yorta Yorta, they have been seeking their authenticities through making the "Land Use and Occupancy Map" supported by Canadian specialists since 2006. The aims of this map are the following; firstly, it lets Australian society acknowledge the Yorta Yorta as Traditional Owners on the river and forest; second it delivers social, economic, environmental and cultural outcomes for the Yorta Yorta and; third it gives the Yorta Yorta opportunity to get partnerships based on respect, honesty, and capacity to participate equally, with shared responsibility and clearly defined accountability and authority (see map in **Figure 12**). The Yorta Yorta collected around 6600 sites as their cultural and spiritual significance from 66 respondents and they trained six coordinators for cultural map who have responsibility to record their traditional and living knowledge of forest and river into the map and take the living knowledge over the next generations. The use of this information will be utilized by the YYNAC to negotiate outcomes in relation to natural resource management, in particular environ-

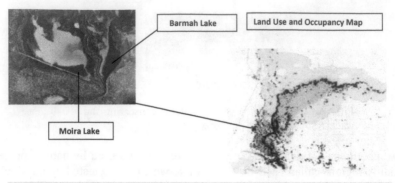

Figure 12. Barmah forest as human body.

mental management plans, watering plans with a view to establish discussions with government regarding Yorta Yorta's right to have allocations of cultural water to return water flow to the areas that need it the most to encourage the spawning of fish and other species and therefore ensure their existence for future generations of Yorta Yorta people.

7. Conclusions

In this paper, an environmental anthropological study of watershed management regarding water quality conservation of forest as catchment area in the southern part of Australia was conducted. As a result on the human activity contributed to the water quality conservation, the importance of constructing FUBEN-EKI (Further Benefit of a Kind of Inconvenience) environment-social systems and the management of water resource based on the slow sand filtration imitated the filtration of forest were recognized respectively. The concluding remarks are as follows: 1) The proposal toward a water quality conservation of watershed/forest as a catchment area is as follows; In the water quality conservation, the value change/consciousness change for water are needed. For that the recognition of the three factors (visualization, awareness and circulation of connection) is very important for water. "Visualization of water quality" as an incentive for the effort of the conservation of catchment area is effective. In concrete, we consider that the mapping of the catchment area water quality by the sounds is appropriate; 2) And the approach required in contrast with the independent effort of water quality conservation driven by the watershed community residents is as follows; eco-money (environmental tax) is effective as a standard currency toward bi-directional connection between man and the forest. Using this money, it is possible to construct the values-driven management. Regarding the introduction and its use, we think that the planning of the environment investment strategy (the practical business administration of the environment) for which becomes a mortgage of sustainability is very important.

Acknowledgements

We acknowledge the support for the research in this paper by the Yorta Yortanation Aboriginal cooperation and The Goulburn Broken Catchment Management Authority, Melbourne, Australia. We also thank insightful comment for our research by Neville Atkinson and Lee Joachim working at the Authority and Nicole Freeman working at St Augustine's P-12 College, Melbourne, Australia. Furthermore, the authors are grateful to Dr. Dhundi Raj Pathak for his comments on this paper.

References

[1]   Ministry of Environment (2014) Annual Report on the Environment in Japan.

[2]   Hiratsuka, A., Shigemitsu, S. and Murota, A. (1996) Study on Water Environmental Management for an Urban Mountain Hill Catchment. *Journal of Osaka Sangyo University, Natural Sciences*, No. 100, 205-214.

[3]   Fukuzawa, T. (2014) Jyosui Shori Sisutemu no Genjyo to Kadai (Current Situations and Issues of Water Treatment System). Graduation Thesis, Osaka Sangyo University, Osaka.

[4]   Tomonaga, Y. (2013) Ostraria Senjyumin no Tochiken to Kankyokanri (Australian Aboriginal Land Rights and Environmental Management). Series of International Human Right Issue 84, Akashi Shoten, Tokyo.

[5]   Curr, E.M. (1965) Recollections of Squatting in Victoria from 1841 to 1851. Melbourne University Press, Melbourne.

[6]   Fahey, C. (1986) The Barmah Forest: A History. The Department of Conservation, Forests and Lands, Vic.

[7]   Weir, J. (2009) River Country: An Ecological Dialogue with Traditional Owners. Aboriginal Studies Press, Canberra.

[8]   Clode, D. (2006) As If for a Thousand Years: A History of Victoria's Land Conservation and Environment Conservation Councils. The Victorian Environment Assessment Council, Melbourne.

[9]   Goulburn Broken Catchment Management Authority (2012) Goulburn River Water Monitoring Report 1995-2010. Goulburn Broken Catchment Management Authority, Shepparton, Vic.

[10] Do.

[11] Hiratsuka, A. and Pathak, D.R. (2013) Application of Ultrasonic Waves for the Improvement of Water Treatment. *Journal of Water Resource and Protection*, **15**, 604-610. http://dx.doi.org/10.4236/jwarp.2013.56061

[12] Hiratsuka, A., Tomonaga, Y., Yasuda, Y. and Tsujino, R. (2014) Improvement of Water and Wastewater Treatment Process Using Various Sound Waves—A Consideration from the Viewpoint of Frequency. *Journal of Water Resource and Protection*, **6**, 1464-1474. http://dx.doi.org/10.4236/jwarp.2014.615135

[13] Kida, K. (2014) Utsukushii Nihon no Shizen-on CD Book (Beautiful Japanese Natural sound CD Book) (in Japanese). Mukku, 27.

[14] Kawakami, H. (2011) Fubenkara Umareru Dezain (Design Resulted from Inconvenience). Dojin Sensho, 10-18.

[15] Hiratsuka, A., Wakae, K. and Yasuda, Y. (2010) An Existential Lifestyle in Our Global Environmental Age—From Homo-Economics to Homo-Environmentics. Heiwagaku Ronshuu IV (*Osaka Sangyo University Journal* IV), Sanken Sousho 32 (Series of the Institute for Industrial Research of Osaka Sangyo University, 32), 143-157.

[16] Sueishi, T. (1996) Kankyojinbungaku no furontia (Frontier of Environmental Literature). In: Society for Environmental Economics and Policy Studies, Eds., *Kankyokeizai-Seisakukenkyu no Furontia* (The First Number), Toyo Keizai Inc., Tokyo, 160-169.

[17] Miyamoto, K. (1996) Ijikanona Shakaini Mukatte (Toward to Sustainable Society). Iwanami Shoten, Tokyo, 123-128.

[18] (2013) A Poem Written by Ho TO, Shunbo 712-770.

[19] Nakamoto, N. (2008) OishiiMizu no Tsukurikata—Seibutsu Jyokahou (Safe Water is Everyone's Right—A Manual on Slow Sand Filtration, An Ecological Water Purification System). TsukijiShokan, Tokyo.

[20] Nihon no Mizu wo Kireinisuru Kai, Ed. (2009) Heiseino Meisui Hyakusen (Selected One Hundred Exquisite Waters of Japan). Gyosei Shuppan, Tokyo.

[21] Kyu, I. (1976) Toward Water Reuse: Learn It from Your Own Body.

13

Feasibility Analysis of MERIS as a Tool for Monitoring Lake Guiers (Senegal) Water Quality

Seybatou Diop[1], Souléye Wade[1], Moshood N. Tijani[2]

[1]Institute of Earth Sciences, Faculty of Sciences and Techniques, Cheikh Anta Diop University of Dakar, Dakar, Senegal
[2]Department of Geology, University of Ibadan, Ibadan, Nigeria
Email: seybdiop@yahoo.fr, wadesouleye@yahoo.fr, tmoshood@yahoo.com

Abstract

ENVISAT/MERIS scenes of Lake Guiers covering the period 2003-2010 were processed for concentration retrieval of chlorophyll a (CHLa), suspended particulate matter (SPM) and colored fraction of dissolved organic matter (CDOM), *i.e.* the three main parameters relevant to the water quality management of the lake. Estimates in the range of 30 - 117 μg CHLa L^{-1} (average 62.13 μg·L^{-1}), 0.10 - 29.0 mg SPM L^{-1} (average 22.01 mg·L^{-1}), and 1.10 - 1.90 CDOM m^{-1} (average 1.33 m^{-1}) were recorded, suggesting the possibility of occasional poor quality waters in some compartments of the lake. The values calculated as part of this study are consistent with literature data. On the basis of these estimates, interpretations were made as to the feasibility of applying MERIS data for synoptic environmental monitoring purposes. The data were subjected to statistical analysis, including regression analysis and significance tests. Estimates of CHLa and CDOM revealed some level of correlation, which suggests that phytoplankton biomass degradation may account for nearly 47% of the dissolved optical compounds CDOM. Notable areas of high CHLa and CDOM concentrations are found in the southern inshore zone, an environment with less water agitation. In contrast, SPM concentrations tend to increase in environments of very shallow water marked by high water turbulence and bottom mobility. However, it was not possible to fully assess the model performance and detection accuracy of the results due to lack of ground truths. Nonetheless, the results show concentrations that compared well with the *in-situ* data from earlier studies and data reported elsewhere from other lacustrine systems. Therefore, it can be inferred from this study that MERIS data present a useful low-cost (*i.e.* cost effective and readily available) approach for environmental monitoring of Lake Guiers waters with excellent spatial coverage. In addition, the study highlighted the minimal effect of the so-called "bottom effect" on model predictions, despite the small depth of the lake.

Keywords

Lake Guiers, MERIS, Environmental Monitoring, Water Quality, Chlorophyll a, Suspended Particulate Matter, Colored Dissolved Organic Matter

---

1. Introduction

Considerable interest has been focused in recent years on the role of satellite remote sensing data for surface water mapping and their application in lake-water quality monitoring ([1]-[3]). Basically, the procedure relies on analyzing the optical characteristics of solar radiation (*i.e.* the so-called "*water-leaving radiance*") reflected by the open-surface water body under consideration, based on its "spectral characteristics" (*i.e.*, absorption, scattering and other relevant physical laws). The conceptual model exploits the causal links between the optical properties of the sensed signal and the inherent optical properties (IOPs) of the individual water-constituents. In practice, however, several additional factors present within the instantaneous field-of-view (IFOV) of the sensor (e.g., aerosols, atmospheric water vapor and gases) influence the satellite sensed (bulk) signal, which typically contains spectral contributions from these materials. These factors usually compounded the measurements and data interpretation. Therefore methods of signal correction and processing needed to be applied to the sensed signal to remove those disturbances in order to allow for the retrieval (extraction) of genuine ground spectra from the target object.

During recent years there has been a great deal of research devoted to building up new algorithms appropriate to overcome image processing difficulties and/or reconstruct the sensed signal ([4] [5]). Algorithms are now available that can be used for the extraction of lake water quality parameters (WQPs) from satellite images. Such algorithms are described in several publications (e.g., [6]-[9]). Several applications can be cited that have demonstrated their usability: [1] [6] [10]-[14].

This study was part of a research project aimed at developing a remote sensing approach towards the implementation of a GIS decision support system for the lake water management. The investigated WQPs are the three key model components for coastal waters ([15] [16]): 1) Chlorophyll-a (CHLa) [referred to as an index for phytoplankton biomass development]; 2) total suspended particulate matter (SPM) [taken as the dry weight of all suspended inorganic particles] and 3) colored fraction of dissolved organic matters (CDOM) [so-called "*yellow substances*", consisting of aquatic humic and fulvic compounds, the presence of which has often been associated with acidic waters].

This paper reports on the preliminary results of investigations on the applicability of the commercially available MERIS image data for concentration retrieval of WQPs for Lake Guiers. The aim was primarily to employ this method as a reconnaissance tool for water quality monitoring and mapping of this shallow lake, using the FUB/WeW processor developed by ([14]). To date, research using this approach has focused essentially on lakes with greater extension and depth. Our specific objectives were twofold: firstly, the evaluation of performance of the FUB/WeW algorithm for signal correction of the raw MERIS scene (1b-image product level) and quantitative estimations of WQPs for the study lake; and secondly, to assess the variations in temporal features of the water quality in the lake and its dynamics based on a time series of WQPs estimates derived from the MERIS 2b-image end-product level. However, due to a lack of ground truths, background data from early studies were used to aid interpretation and tentative validation (*i.e.* calibration) of the derived WQPs data in this study.

2. Theoretical Concept

The conceptual model of the technique employed in this study relies on the relationship between the reflectance (*R*) recorded on board the MERIS-sensor and the inherent optical properties (*i.e.* IOPs or spectral characteristics) of the sensed open-water body. This relationship can be put in the form (e. g., [17]):

$$R(\lambda) = \gamma \cdot \frac{b(\lambda)}{a(\lambda) + b(\lambda)} \tag{1}$$

where $R(\lambda)$ is the spectral reflectance measured in band λ. The γ term is dependent on the directional illumination geometry of the light field emerging from the water body and may be expressed as ([18]):

$$\gamma = \frac{1}{1+\dfrac{\overline{\mu_d}(\lambda)}{\overline{\mu_u}(\lambda)}} \tag{2}$$

where-in the term $\dfrac{\overline{\mu_d}(\lambda)}{\overline{\mu_u}(\lambda)}$ represents the ratio of the average cosine of the down-welling (incident) light to that of the up-welling light over the water body.

The term $a(\lambda)$ is the total absorption coefficient of the water body characterizing the decay rate of incident sunlight due to the process of absorption by the water molecules themselves and by the various suspended and dissolved components that have distinctive spectral signatures, and

$b(\lambda)$ is the total scattering coefficient of the water body representing the decay rate of incoming sunlight due to (back)scattering at the surface of these elements.

In Equation (1), each term is followed by the symbol for wavelength in parentheses (λ), since measurements are made in terms of monochromatic components. The components of absorption $a(\lambda)$ and $b(\lambda)$ thus symbolize the pooled spectral effects of both the water molecules and all water resident constituents on the incident sunlight reaching the target surface water body. Obviously, since the spectral properties of a reflecting medium relate to the specific inherent optical properties (IOPs) of its individual constituents, the per-unit concentration of the constituents and the coefficients $a(\lambda)$ and $b(\lambda)$ can be expressed as explicit functions that include the spectral properties of these constituents as follows (conjectural three components model for "*case*-2 *waters*"; ([15] [16]):

$$a(\lambda) = a_w(\lambda) + Pa_p^*(\lambda) + Ya_y^*(\lambda) + Sa_s^*(\lambda) \tag{3}$$

$$b(\lambda) = b_w(\lambda) + Pb_p^*(\lambda) + Sb_s^*(\lambda) \tag{4}$$

where the subscripts w, p, y and s refer to the optically active constituents (OACs) in the considered medium and stand for water, phytoplankton, yellow substances and inorganic suspended particles respectively, while the symbols P, Y and S refer to their respective concentrations.

In the notation used here, the asterisks denote specific coefficients of the components. For instance, $a_p^*(\lambda)$ refers to the specific back-scattering caused by phytoplankton in spectral band λ, while $b_s^*(\lambda)$ is the specific back-scattering coefficient of suspended particulate matter. Pure water forming a constant background optical property, its coefficients are written without asterisk.

3. Environment Setting of the Study Area

Lake Guiers is located in northern Senegal, lying between 15°35' and 16° Long.-W, and 15°50' to 16°35' Lat.-N (**Figure 1**). It is a shallow lake with a north-south elongation (about 50 km long and 2 to 7 km wide), and a maximum depth of 6 m in the central zone. It is an essential water resource for Senegal. It receives drain waters from the adjoining Senegal River at the border of Senegal with the Republic of Mauretania. The lake drains from north to south with maximum depths at the central zone, while the depth is shallower towards the southern end. It is in this inshore zone, which is partially exposed at low water that the lake discharges into the Ferlo valley, acting as the principal surface water outflow in addition to water losses from evaporation or vertical leakage to the underlying aquifer system.

Lake Guiers is bottoming at around −2 m below mean sea level and, at its highest mean water level (*i.e.* +4 m above mean sea level), it may expand to cover a surface area of about 320 km². Especially characteristic is the presence of aquatic plants (such as emerged/fixed and floating macrophytes) that populate the near shoreline (bank margins) environment, an area of very shallow water. These aquatic plants are important because they supply a local stabilization of the lake floor and act as trapping agents that allow suspended sediment to settle on the lake bottom and stabilize as soon as it is deposed. Moreover, they provide low energy wave action and surfaces for attachment or food gathering, and generate a good habitat for benthonic organisms. The offshore environment is the area of deeper water with a maximum water depth ranging from 4 m to 6 m. [19] describe this zone as an environment where uniform ecological conditions prevail which provides an optimum environment for algal growth. There is a tendency for phytoplankton to locally develop from the low-water shoreline environment to the offshore environment. The growth density of the principal living organisms is related to water depth and movement, and their distribution defines three distinct ecological zones (see **Figure 1**), namely

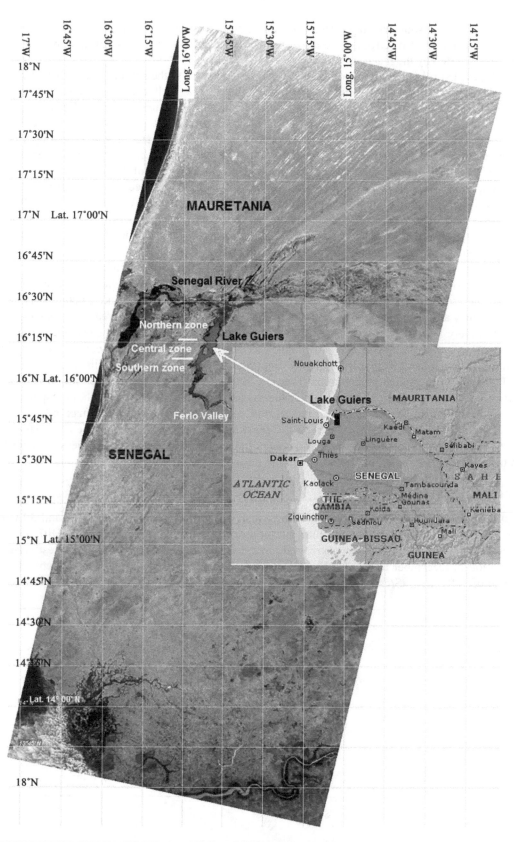

Figure 1. Coverage area of the April 27th, 2007 MERIS FR 1b-composite image showing the location of the study area within Senegal (flow is from east to west for Senegal river and north to south for Lake Guiers).

northern, central and southern zones. However, in the typhae-dominated southern zone and alongside the bank margins, the occurrence of CHLa, in terms of primary production (algal activity), may be low due to the high competition exerted by the macrophytes vegetation.

Initially, the hydrological regime of Lake Guiers was subjected to the natural fluvial dynamics of the Senegal River. Nowadays, the regulating dams constructed on the river, as well as the channelization of the river, make possible a large-scale storage and use of river water independent of natural flooding. Before any regulation was attempted, the hydrology and hydrochemistry of the lake were characterized by major consequences on the hydrophytes settlements, due to the phases of alternating high and low water levels (*i.e.* drought events) and accompanied high salinity. [20] and [21] found that the new hydrological conditions generated by the regulating dams (viz., the stabilization of the lake stage at fairly constant elevation of 2 m above mean sea level (see **Figure 2**) and, hence, the permanent softening of the lake water quality) are ecological factors that encouraged the expansion of an important aquatic vegetation of free floating and fixed macrophyte (essentially *Typha*, *Phragmites*, *Pistia stratiotes*, *Echinochloa* and *Nymphea*). According to [22], this proliferation of aquatic plants best explains the rapid spread of *schistosomiasis*, the larval forms of parasitic blood flukes that currently affect the health of populations living in lakeside villages.

Lake Guiers' main uses are domestic, agricultural, industrial water supply and fisheries. Its water supply accounts for nearly 60% of the daily water consumption of the capital city of Senegal (Dakar) and its suburbs. However, due to its surrounding fertile soils, irrigation agriculture within the lake catchment is increasingly impacting on the water quality, the extent of which is not well documented. The area is characterized by the extension of relatively flat and open landscapes of SW-NE trending sand dune systems (*i.e.* Quaternary ergs) overlying the Tertiary substratum. The lake bed geology, as described from a series of cores to a depth of 5 m, grades downwards from bluish-green colored silts to fine sand sediments ([23]). The climate is arid (Sahelian type), with almost all of the annual precipitation (average: 300 - 400 mm/yr) occurring during the rainy season which lasts from July to September. Vegetation is of the savannah type, namely tiger bush, interspersed with bare soil and agricultural fields. A mean annual air temperature of about 26.3°C (min: 12°C, max: 42°C), with small deviations in monthly mean temperature, reflects the proximity of the sea. The amplitude of variations does not exceed ±5°C, with the hottest months being from May to October. The area mostly experiences north-easterly Saharan trade winds (Harmatan) of 2 - 6 m/s monthly mean.

4. Background Information

The lake waters are of pH 6.8 to 9 and vary seasonally in temperature between 18°C and 30°C (with a mean 25°C). Amplitudes of the daily temperature variations of the waters rarely exceed 5°C. Differences in temperature between the top and bottom water layers are in range 4°C - 5°C or less, with the greatest variation occurring in the period of December to March according to [24]. Data from this study indicate electrical conductivity in the range of 160 - 210 μS/cm (period 2002-2004), total dissolved solids in the range of 198 - 250 mg/L, with

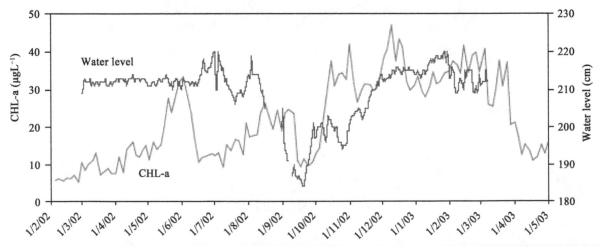

Figure 2. Pattern of variations of Lake Guiers CHLa contents and water level in the observation period February 2002-May 2003 (modified after [19]).

Ca^{2+}, Na^+ and HCO_3^- and Cl^- being the dominant ions, while nutrients PO_4^{3-} and NO_3^- are in the range of 0.2 - 1.8 and 0.2 - 2.5 mg/L respectively. The oxygen saturation typically is above 60%. Under these conditions, the euphotic layer of the lake almost matches the entire water column. Environmentally, this means very uniform ecological conditions which provide ideal circumstances for phytoplankton development.

Up till now, the only literature available on the lake's water quality generally focused on the occurrence of CHLa and SPM (e.g., see [19] [24] [25]). However, little is known about the occurrence of CDOM. [19] measured CHLa contents in the range of 5 to 47 µg CHLa L^{-1} and suggested a sort of seasonal trends (as presented in **Figure 2**) which might be related to phytoplankton biomass variations. However, this may not be considered a regular/cyclic pattern, as evident from the comparison of CHLa occurrence for the periods of February-April, 2002 and February-April, 2003. Similar results was obtained during the winter months by [25], who noted the occurrence of higher CHLa contents (45 - 70 µg CHLa L^{-1}) as an obvious and characteristic effect of temporal phytoplankton variations.

A similar scenario was observed for SPM concentrations by [25], with values in the range 15 - 45 mg·L^{-1} (mean 29 mg·L^{-1}). His **Figure 3** ([25]) presents evidence of a declining water turbidity during the period May-September 2005. SPM values are lowest in the period of July-August, a trend that is consistent with the seasonality of the climate as discussed further below. SPM concentrations increase steadily for the rest of the year to reach the maximum peak in February-March. The SPM values of [25] are somewhat higher than those reported by [26] along an N-S transects of the lake, with values of 3 to 30 mg SPM L^{-1} (average 13 mg SPM L^{-1}). The [26] measurements were lowest in the northern ecological zone, with values of 3 to 10 mg SPM L^{-1}, and which increase towards the thicker central portion of the lake, reaching values of up to 30 mg SPM L^{-1}. Toward the south, the SPM concentration values decrease again, a situation that can be attributed to the presence of macrophytic vegetation that incidentally decreases the hydrodynamic conditions and thus favoring particle sedimentation. However, SPM concentrations as high as 30 mg SPM L^{-1} may occur also in the southern shallow water zone of the lake as well, as observed by [26] at downstream section of the lake, near Keur Momar Sarr embankment. The probable explanation is a sporadic change in turbidity due to water agitation or particles input from desert sand storms. Apparently, some patterns of spatial and temporal variations of the SPM concentrations were suggested by the various authors, which probably indicate some characteristic features that typify the lake behavior. Evidence comes from the fact that the SPM load of the lake usually decreases during the summer months with the onset of the rainy season. This is as a result of the episodic effect brought about by the seasonal changes in wind direction, as monsoon airstreams blow onto the land, thus supplying the trocken Saharan trade winds (harmattans). In the seasonally warm periods (*i.e.*, mostly during the months of February and March) sand particles input into the lake waters is created by the large dust storms from the Sahara Desert. Strong winds that carry desert dusts into the atmosphere are consistently produced every year as the air over the hot desert regions is heated and expands in volume causing it to rise upwards.

Another important feature of the lake waters reported by [24] during the period 2002 to 2005 is the presence of high phytoplankton biomass (with a pluriannual average value of about 46 µg CHLa L^{-1}) that immobilizes most of the nutrients in the lake. According to the author, cyanophytes and chlorophytes dominated the phytoplankton community of the lake during the period, with yearly averaged biomass productivity as 6.5 mg C (mg

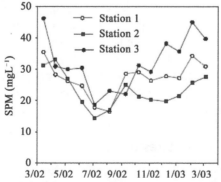

Figure 3. Pattern of variations through time of SPM concentrations in Lake Guiers for the observation period from March 2002 through March 2003. Note: SPM concentrations shows similar variation at the three stations, with values in range of 15 - 45 mg·L^{-1} ([25]).

Chla)$^{-1}$·h^{-1}. During this observation period, the CHLa content of the lake waters varied with yearly averages of about 20 μg CHLa L^{-1} in 2002, 18 μg CHLa L^{-1} in 2003, 28 μg CHLa L^{-1} in 2004 and 51 μg CHLa L^{-1} in 2005.

Therefore, judging from the broad range of CHLa values (from 5 to 70 μg CHLa L^{-1}) reported by the various researchers ([19] [24] [25]), one may conclude that Lake Guiers' trophic index may vary appreciably, almost from oligotrophic to eutropic state. It must be emphasized that the applied margin of safety for lake water is CHLa concentrations about ≤9 - 10 μg·L^{-1}; this is because taste and odor problems begin once the chlorophyll-a concentrations reach higher values. For biota and others hydrophobic organic substances in lake waters, the EU fresh water quality standards with reference to SPM is 15 mg/l.

5. Database and Methodology

MERIS products are satellite image data acquired by the Medium Resolution Imaging Spectrometer (MERIS) on board the European Space Agency (ESA) Environmental Satellite (ENVISAT)-1 launched on March 2002. MERIS is a typical nadir-looking push-broom imaging system that enables monitoring of the Earth's atmosphere and surface with a repetition time of 3-day intervals. It can be operated either in direct mode (so as to deliver a full-resolution-FR product of 300 m on-ground resolution) or in averaging mode (so as to produce a reduced resolution-RR product of 1200 m on-ground resolution). The MERIS sensor operates with 15 programmable bands in the spectral range 412.5 - 900 nm, with a 10 nm average bandwidth. Further details on the instrument are reported in [27].

The MERIS data used in this study were provided by the TIGER project N° 2793 on behalf of the ESA, and covers the observation period from 2003 to 2010. Most observations were concentrated on the months of January, March, April, May, August, October and November. The average monthly mean concentrations were then calculated for each estimated parameter based on the sensor's effective repetition rate of 3-day intervals (*i.e.* using the 9 - 10 scenes or so taken during the month). The calculated monthly mean concentrations were subsequently averaged per annum, and then averaged over the observation period (*i.e.* 2003-2010) to yield the long-term (pluriannual) mean for each parameter.

The studied scenes comprised a set of rough MERIS full resolution 1b-image product-levels and another set of 2b-image end-product levels. The focus of the methodological approach was:

- Objective 1: to test the suitability of the FUB/WeW algorithm ([14]) for signal correction and concentration estimates and mapping of the lake WQPs using the rough 1b-image product, and
- Objective 2: to demonstrate the potential use of the 2b-end-products (*i.e.* on board ENVISAT-1 already processed images) for changes detection.

Objective 1 was addressed by selecting the image presented in **Figure 1**. This image taken on April 27th, 2007 was the only cloud-free scene covering the study area out of a set of ten 1b-scenes available for purchase in Phase I of this research project; the remaining scenes had intolerable levels of cloud cover. The processing method using the MERIS FR 1b-product as input for the FUB/WeW algorithm has been presented elsewhere ([28]).

In contrast, cloud cover was not a primary limitation for Objective 2, which was an application based on the standard MERIS FR 2b-end product level. These are "*ready-to-use*" image data that offer the opportunity of straightforward visualization of the generated WQPs using the BEAM (Basic ERS & ENVISAT (A)ATSR and MERIS)-VISAT software. The BEAM-VISAT Toolbox developed by Brockmann Consult for ESA (see [29] for more details) enables viewing of the output quality products (*i.e.* physical parameters) contained in the level 2b-product. The BEAM-VISAT version used here is Version 4.8 (http://www.brockmann-consult.de).

Finally, it should be pointed out that surface water quality mapping based on MERIS FR images typically represents a ground-based sampling on a 300m-by-300m grid cell basis as the smallest unit area (namely, the image pixel size). Each grid cell (or pixel) is assigned a parameter value averaged out over its area.

6. Results and Discussion

6.1. Description of the FUB Model Results

Table 1 presents the summary statistics of the pixel values for the three WQPs derived from the April 27th, 2007 MERIS FR 1b-product level by applying the FUB algorithm, while **Figure 4** shows their frequency distributions (FD) along with fitted normal distribution model curves N[μ, σ], wherein the symbols μ and σ represent, respectively, the mean and standard deviation of a given parameter respectively.

Table 1. Statistical summary of the retrieved concentrations values for the three analyzed WQPs.

Parameter	SPM (mg·L^{-1})	CDOM (m^{-1})	CHLa (µg·L^{-1})
Sample size	1151	1147	877
Mean (average)	22.03	1.33	62.13
95% conf. int. (mean)	±1.4%	±1.1%	±1.3%
Median	22.99	1.31	60.54
Mode	22.81	1.29	58.32
Geometric mean	19.79	1.33	60.96
Variance	28.07	0.01	158.85
Standard deviation	5.3	0.11	12.6
Standard error	0.16	0	0.43
Minimum	0.17	1.1	32.36
Maximum	28.92	1.88	117.58
Range	28.75	0.77	85.23
Lower quartile	22.18	1.27	55.24
Upper quartile	23.79	1.37	67.2
Skewness	-2.78	1.94	1.28
Coeff. of variation	24.04	8.53	20.29

As shown in **Table 1**, the respective mean, median and mode for SPM, CDOM and CHLa are close enough as to reasonably corroborate the standard assumption for normal distribution. Moreover, the statistics indicate a 95% confidence interval on mean concentration values in extremely narrow ranges (*i.e.* from ±1.1% to ±1.4%), thus suggesting a high probability for an efficient estimate of the hypothetical "population mean" for each parameter.

Close inspection of the range of mean values reported in **Table 1** reveals that the average concentrations of all three WQPs exceed the limits of drinking water regulations, suggesting poor quality waters within the lake compartments at the time of image acquisition. CDOM values have a moderately lower coefficient of variation (8.53%), compared with CHLa and SPM values with 20.3% and 24%, respectively; this can be attributed to the fact that CDOM are miscible constituents and thus may be distributed homogenously within the lake waters. However, CHLa and SPM constituents are present in the water medium as suspensions and, therefore, may exhibit higher spatial variability due to the complexity of the lake ecosystem in terms of structure and physiography.

6.1.1. Comparative Assessment

CHLa values: At least two major peaks (one at around 62 µg CHLa L^{-1} and another at about 72 µg CHLa L^{-1}) can be noticed in the frequency distribution (FD) of CHLa (**Figure 4(a)**), which indicates mixed feature or spatial variability of the CHLa concentrations. CHLa values inferred for pixels located along the bank margins are fairly high (>80 µg CHLa L^{-1}), resulting in the chlorophyll pigment production from the macrophytes vegetation populating the near-shore zones of the lake. Moreover, the 95% confidence interval, involving the mean for CHLa, from 61.32 to 62.94 µg CHLa L^{-1}, is within narrow limits. This means that if the retrieved CHLa values are averaged to symbolize the standard trophic state of the lake at the time of satellite overpass, then the lake status can be deemed "eutrophic" with a 95% probability at that moment. This is consistent with field observations from early studies ([19] [24] [25]).

SPM values: The frequency curve for SPM (**Figure 4(b)**) exhibits two dominant peaks, the first at about 25 mg·L^{-1}, and the second at about 29 mg·L^{-1}. This also suggests a mixed feature and variability in the turbidity level of the lake water. However, the distribution pattern reveals that SPM values have roughly three population groupings. The fitted normal distribution curve is skewed to the left indicating that most SPM values are "larger" than the mean value of 22.03 mg·L^{-1}. This means highly turbid conditions throughout most parts of the lake, a finding that is consistent with similar results previously obtained by other investigators (e.g., [25]). It could be

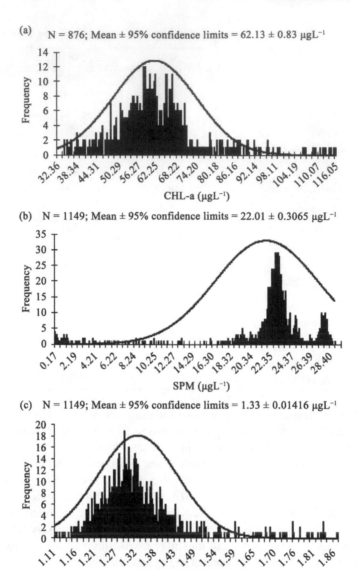

Figure 4. Frequency histograms with fitted normal distribution curve for (a) CHLa values (b) SPM values and (c) CDOM values.

inferred that the apparently high turbidity level of the lake could be the result of a variety of factors, including, for example: sand particles input from Saharan dust storms that are predominantly active during the period of the year in focus (April month); contributions from feeding river waters (as the sediment-laden streams water flow into the lake); and the effect of high-speed Saharan trade winds that typically induce wave motions and bottom particle re-suspension within shallow water depths.

CDOM values: The histogram plot of the CDOM concentration values (**Figure 4(c)**) indicates a fairly unimodal distribution due to a smaller variance (see **Table 1**). There is no literature data on CDOM available for making comparisons with the FUB-predicted values. However, the observed lower range of 0.77 CDOM m^{-1} (with measurements ranging from 1.105 to 1.877 CDOM m^{-1}) is consistent with similar narrow ranges and orders of magnitudes reported in the hydrological literature ([30] [31]).

As part of further data evaluation, the studied WQPs are expressed in terms of contoured concentration maps as presented in **Figure 5** where the variability in colour is related to concentration variations. Visual inspection and assessment of the spatial variability revealed a general westward increase in CHLa and CDOM concentrations as against the SPM concentrations. High CHLa and CDOM concentrations and low SPM concentrations in the western shoreline are thought to result from the occurrence of the macrophytic vegetation on the lake reaches.

Moreover, the highest SPM concentrations are found upstream (near the Mbane village) and downstream of the lake (near the Sear village). Compared to the central zone, these are locations where the lake is shallower and/or narrower, *i.e.* places where the lake hydrodynamics, and wind-induced particle re-suspension effects, are more important. The spatial variability in the SPM concentrations observable in **Figure 5(c)** in the N-S direction of the lake is consistent with the one depicted in previous studies ([24] [26]).

6.1.2. Correlation Analysis

Figure 6 shows a tentative correlation between FUB-predicted SPM values and early *in-situ* measures compiled from the literature ([26]). With knowledge of the geographic coordinates of the locations of these field measurements, it was possible to locate and spatially match-up the *in-situ* measurements with SPM values retrieved from the MERIS 1b-image at corresponding pixel positions.

The best correlation coefficient, r = 0.741 (R^2 = 0.5491) obtained by discarding one outlier data point, is not statistically significant at the 95% confidence level (ANOVA P-value = 0.0567), suggesting a poor linear

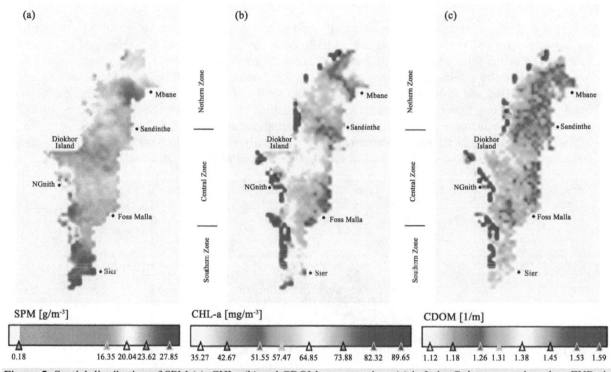

Figure 5. Spatial distribution of SPM (a), CHLa (b) and CDOM concentrations (c) in Lake Guiers waters based on FUB algorithm retrieval from the April 27th, 2007 MERIS observation.

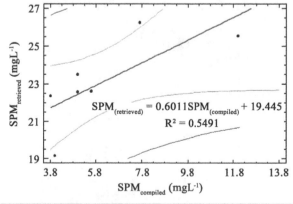

Figure 6. Plots of FUB-predicted SPM values versus SPM values measured by ([15]).

relationship between the two data sets. However, nearly 55% of the variance of FUB-derived SPM values could be explained by the fitted line, judging from the coefficient of determination R^2. This result increases the confidence level in the reproducibility of the data. Nonetheless, it is possible to infer that real-case regression equations derived from comparing FUB retrievals to ground truths might provide a basis for reliable WQPs estimates.

In addition, assessment of the possible relationship between CHLa and CDOM concentrations as presented in **Figure 7**, revealed a correlation coefficient (r) of 0.686 ($R^2 = 0.4709$), which is statistically significant at 0.05 level (ANOVA P-value = 0.0000). This signifies that nearly half of the variance of the CDOM absorption can be explained by the variance of the active phytoplankton biomass. Presumably, the implication is that the degradation of phytoplankton cells within the system accounts for a CDOM proportion of 47%.

However, the remaining 53% proportion of CDOM that cannot be explained by the fitted equation may be related to an "*allochthonous*" source which, in turn, may explain the markedly high CDOM contents in the lake. Other factors, which may influence the CDOM, include natural cycling of elements (through the decay of plants and phytoplankton) and direct anthropogenic activities such as agricultural practices and industrial wastewater inputs.

Figure 8 displays the correlation between CHLa and SPM values with a huge point scattering with respect to the best-fitted line (SPM = −0.1729 CHLa + 32.059). Obviously, about 18% of the variance of SPM values can be explained by the fitted equation, with a correlation coefficient (r) of 0.425. This is a poor negative correlation which is statistically significant at the 5% level (ANOVA P-value = 0.0000), suggesting possible inverse relation between CHLa and SPM values. High CHLa values (>70 µg CHLa·L^{-1}) basically outweigh the slope of the fitted regression line. Nonetheless, a closer visual examination of the plot revealed that SPM concentrations are generally found to decrease with increasing CHLa concentrations under two different scenarios:

Scenario A: At higher CHLa concentrations, which typically indicate chlorophyll pigment production from macrophyte vegetation, the apparent inverse relationship with SPM can be attributed to the reduced hydrody-

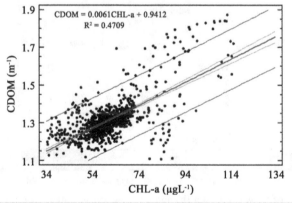

Figure 7. Relationship between CDOM concentration and CHLa concentration.

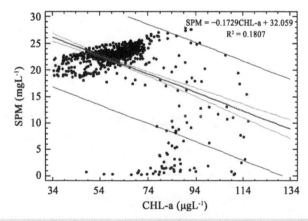

Figure 8. Relationship between SPM concentration and CHLa concentration.

namic conditions, which favour sedimentation of suspended particles under such macrophyte-dominated subsystems. This is consistent with the low SPM concentrations in places with dense aquatic macrophytes development as reported in [26].

Scenario B: In macrophyte-dominated subsystems, the fact that phytoplankton is of minor importance, as reported in [31], clearly supports the higher CHLa values (>70 µg CHLa L^{-1}). Thus, this explains the relatively weak positive correlation of the data points with SPM at much lower CHLa values.

6.2. Change-Detection Study

Understanding the lake's dynamics is an important issue for the lake waters management. However, it was not possible to critically examine this issue in this study due to lack of background information. In fact, the methodology requires the need to address the driving forces that affect phytoplankton growth rate that might have an impact on CHLa concentrations. The approach should also take account of the different causes of lake water turbidity, as well as the sources and mechanism of particles input. Many eutrophication models ([32] and [33]) have demonstrated the role of water temperature, intensity of sunlight and nutrients dynamics for phytoplankton growth. However, a high growth level also requires reduced water movement (*i.e.* quiet flow conditions). This is because under high flow, a turbulent hydrologic regime may develop, resulting in turbid waters and strong bottom-water currents that fall into intolerable limits for algal biomass development. Turbidity mainly comes from feeding runoff river water streams that carry large amounts of bed- and suspended-load solids into the lake. This may also possibly originate from sand particles introduced to the lake by Saharan sand storms and particles resuspension due to bottom currents. As part of Objective 2 the following tentative interpretation was made from a chronological series of MERIS 2b-image products, based on these sources.

Long-range forecasting: Long-term mean-concentration values that would provide indications on the lake dynamics over time are presented in **Figure 9**. They range from 2.22 to 23.7 mg·L^{-1} for SPM, 0.134 to 29.3 µg·L^{-1} for CHLa and 0.081 to 0.447 m^{-1} for CDOM. These values are within the range of those sampled from the MERIS 1b-product level. Based on the differences in pixel color, it is possible to suggest that SPM concentrations are highest where fairly high energy conditions prevail (**Figure 9(a)**), viz., in the feed-water entrance of the northern zone, along the eastern near shoreline bank, and the central off-shore zone. These are places where turbidity currents and bottom currents dominate the lake hydrodynamics. In contrast, the lower SPM estimates obtained downstream, in the barrier islands protected southern inshore zone and along the western margins, are attributed to the possible influence of the vegetation. Indeed, the aquatic plants invading these shallow water environments decrease the hydrodynamic conditions, thus allowing suspended sediment to settle. These low energy environments where algal growth might be favoured, are seen (**Figure 9(b)** and **Figure 9(c)**) to exhibit the highest CHLa and CDOM values, yielding maximum pluriannual means in the ranges 23 - 29 µg·L^{-1} for CHLa and 0.294 - 0.447 m^{-1} for CDOM. The correlation between these two parameters is reflected in the similarity of their geographical distribution pattern as shown in **Figure 9(b)** and **Figure 9(c)**.

When examined in detail, algal production occurs mainly on the inner zone and on the barrier island protected southern ecologic zone (**Figure 9(b)**). As explained earlier, the high CHLa level in these environments, and its correlation with CDOM level, is believed to be partly related to the parallel process of biomass reduction (mineralization) as algae decay and, to a certain extent, other processes like plant decay or anthropogenic activities.

Seasonal forecasting: Assuming that changes in the WQPs concentrations come from temporal changes in the driving environmental factors, a change-detection was assessed for the period 2003 to 2010 by comparing monthly trends. In this respect, **Figures 10-12** show co-extend WQP concentration maps of the monthly mean estimates averaged on the monthly scale for six selected months, starting on January each year, with an intervening interval of one month. The comparison demonstrates the general characteristics of the seasonal variability for each WQP. It can be seen that SPM contents are highest around the month of March (**Figure 10**) when dust storms from the Sahara Desert are active, and during the rainy period (*i.e.* August) as turbid runoff river water, that carries a large amount of bed- and suspended-load, flows into the lake.

As may be seen also from **Figure 11** and **Figure 12**, CHLa and CDOM were estimated to show a similar pattern of seasonal variation within the lake compartments, suggesting a related mechanism of origin. The distribution trend revealed increasing concentration values during the months of March, October and November. However, the month of November exhibited the highest CHLa concentrations, while CDOM revealed highest contents on March. This is the period when SPM values are highest (see **Figure 10**), that is, when turbidity conditions

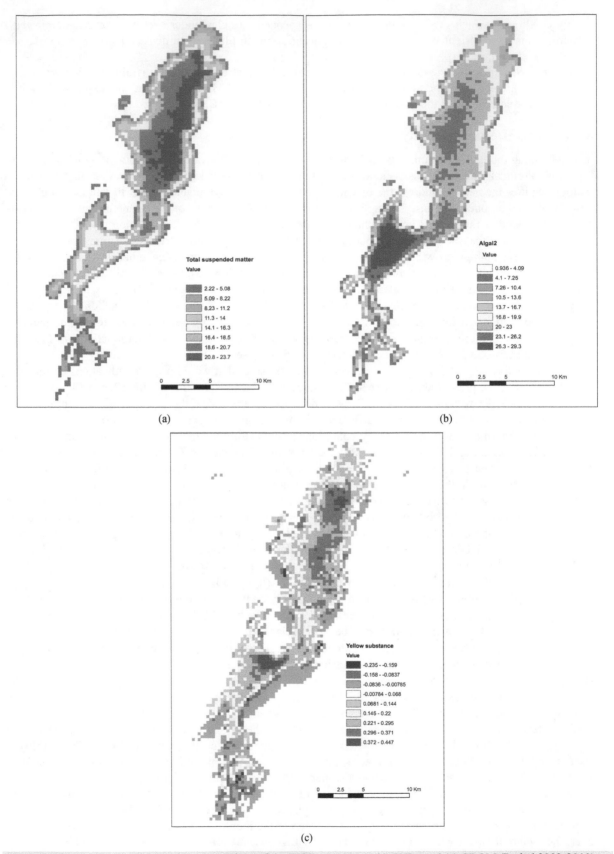

(a)

(b)

(c)

Figure 9. Average pluriannual mean concentrations of the WQPs: (a) SPM; (b) CHLa and (c) CDOM (Period 2003-2010).

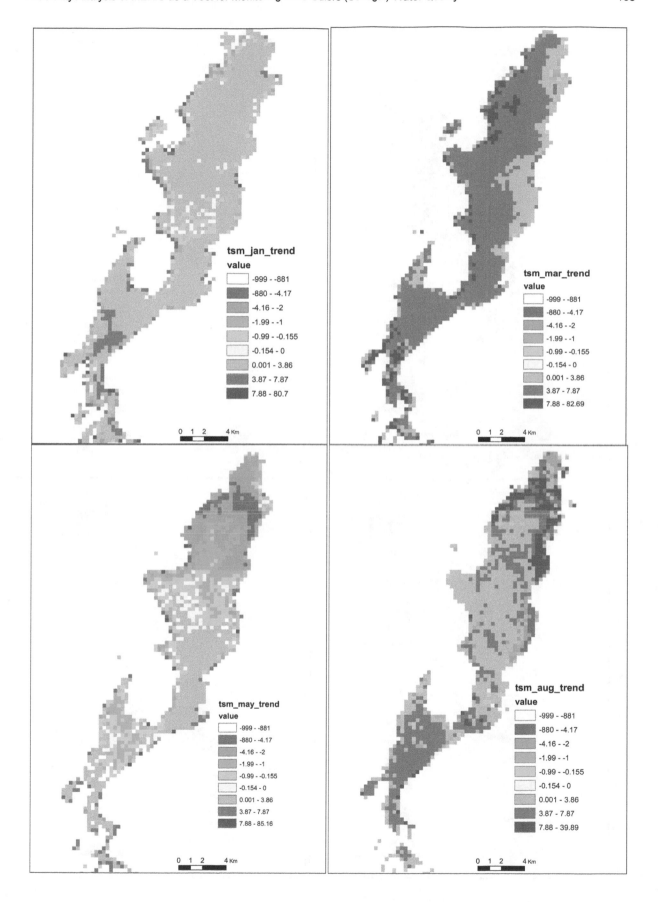

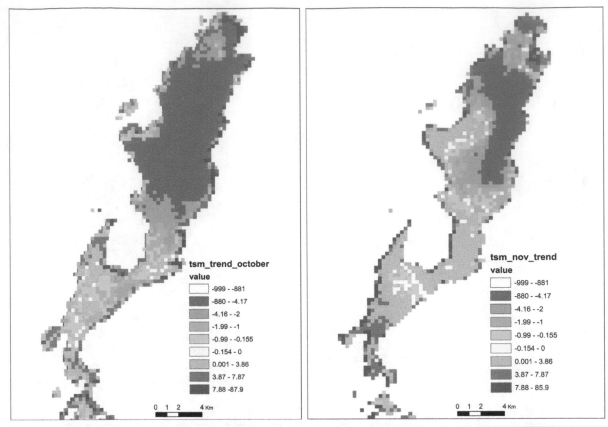

Figure 10. Mean monthly spatial distribution of SPM (period: 2003-2010).

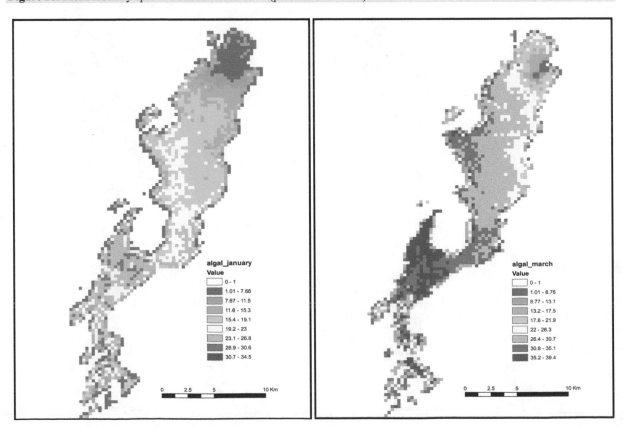

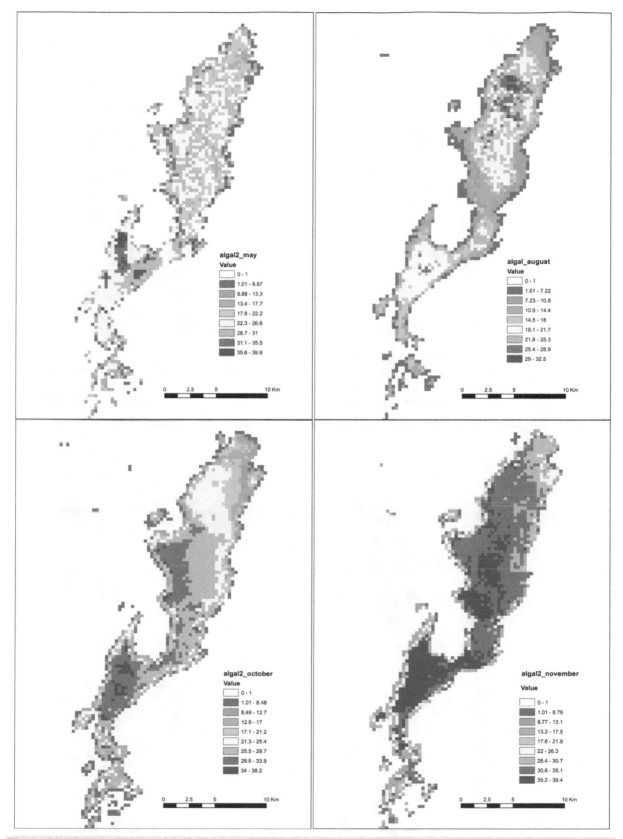

Figure 11. Mean monthly spatial distribution of CHLa (Period 2003-2010). Note: the comparison shows the November month with the highest abundance estimates of phytoplanktonic organisms.

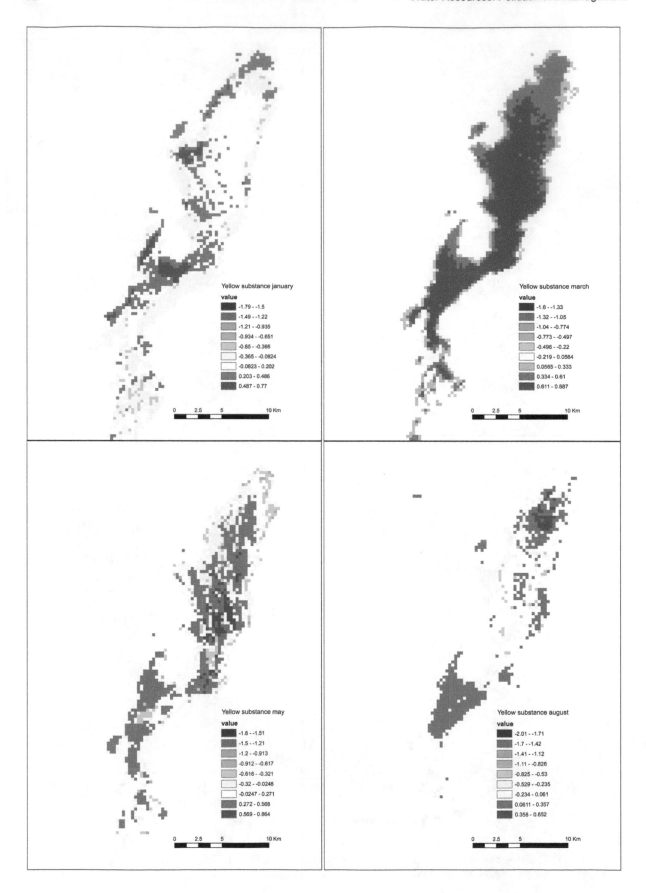

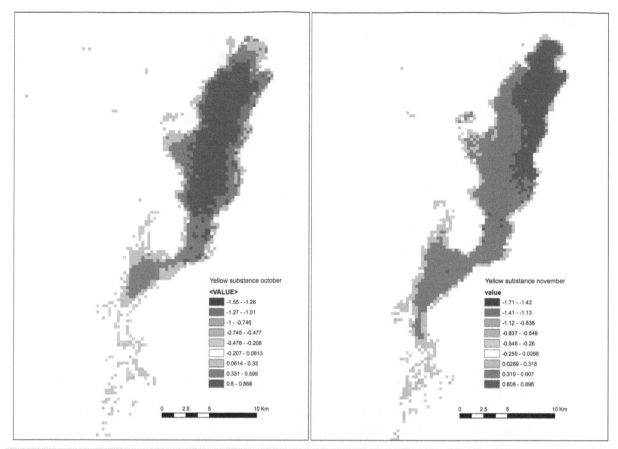

Figure 12. Mean monthly spatial distribution of CDOM (period 2003-2010).

might be environmentally so critical that algae may decompose and result in an abundant presence of CDOM. Nonetheless, the minimum concentration for both CHLa and CDOM in the winter month of January suggests the reduction of algal activity as soon as water temperatures are lower.

7. Summary and Conclusions

This study examined the feasibility of applying MERIS imagery for a synoptic monitoring of the water quality of Lake Guiers, an important surface water resource for Senegal. Based on the application of a chronological data series of MERIS 1b- and 2b-image product levels as input data into the "*easy to use*" BEAM-VISAT software, the concentration of three main parameters relevant for the lake's water quality management was extracted. The values calculated as part of this study matched well with literature data, thus supporting our hypothesis for the capability of this method to provide reasonably valid estimates. Moreover, some characteristic seasonal changes that occur in the lake were recognized and then discussed based on background information. However, an effective assessment of the accuracy of such estimates, and validation of the applicability of the method, warrants the need for ground confirmation through field measurements.

By and large, one of the significant outcomes of this study is the fact that the so-called "*bottom effects*" were proved to be of no relevance in the model predictions, despite the shallow depth of the study lake, bearing in mind that previous studies using this approach were essentially limited to larger lakes with greater depth.

Acknowledgements

The European Space Agency ESA is gratefully acknowledged for the provision of MERIS data within the framework of TIGER/TREES project. The authors wish to thank ITC fellows for their support and contribution, and the invaluable assistance of Dr. Suhyb Salama of ITC (University of Twente, The Netherlands) is particularly appreciated.

References

[1] Dekker, A.G., Vos, R.J. and Peters, S.W.M. (2001) Comparison of Remote Sensing Data, Model Results and *in Situ* Data for Total Suspended Matter (SPM) in the Southern Frisian Lakes. *Science of the Total Environment*, **268**, 197-214. http://dx.doi.org/10.1016/S0048-9697(00)00679-3

[2] Dall'Olmo, G. and Gitelson, A.A. (2006) Effect of Bio-Optical Parameter Variability and Uncertainties in Reflectance Measurements on the Remote Estimation of Chlorophyll-A Concentration in Turbid Productive Waters: Modeling Results. *Applied Optics*, **45**, 3577-3592. http://dx.doi.org/10.1364/AO.45.003577

[3] IOCCG (2006) Remote Sensing of Inherent Optical Properties: Fundamentals, Tests of Algorithms, and Applications. In: Lee, Z.P., Ed., *Reports of the International Ocean-Colour Coordinating Group*, Dartmouth, Vol. 5. http://www.ioccg.org/reports/report5.pdf

[4] Gohin, F., Loyer, S., Lunven, M., Labry, C., Froidefond, J.-M., Delmas, D., *et al.* (2005) Satellite-Derived Parameters for Biological Modeling of Coastal Waters: Illustration over the Eastern Continental Shelf of the Bay of Biscay. *Remote Sensing of environment*, **95**, 29-46. http://dx.doi.org/10.1016/j.rse.2004.11.007

[5] Schalles, J.F. (2006) Optical Remote Sensing Techniques to Estimate Phytoplankton Chlorophyll A Concentrations in Coastal Waters with Varying Suspended Matter and CDOM Concentrations. In: Richardson, L. and Ledrew, E., Eds., *Remote Sensing of Aquatic Coastal Ecosystem Processes*: *Science and Management Applications*, Springer, 27-79. http://dx.doi.org/10.1007/1-4020-3968-9_3

[6] Giardino, C.G., Brando, V.E., Dekker, A.G., Strömbeck, N. and Candiani, G. (2006) Assessment of Water Quality in Lake Garda (Italy) Using Hyperion. *Remote Sensing of Environment*, **109**, 183-195. http://dx.doi.org/10.1016/j.rse.2006.12.017

[7] Schroeder, Th., Schaale, M. and Fischer, J. (2007) Retrieval of Atmospheric and Oceanic Properties from MERIS Measurements: A New Case-2 Water Processor for BEAM. *International Journal of Remote Sensing*, **28**, 5627-5632. http://dx.doi.org/10.1080/01431160701601774

[8] Schroeder, Th. and Fischer, J. (2003) Atmospheric Correction of MERIS Imagery above Case-2 Waters. *Proceedings of the* 2003 *MERIS User Workshop*, ESA ESRIN, Frascati.

[9] Doerffer, R. and Schiller, H. (2007) The MERIS Case 2 Water Algorithm. *International Journal of Remote Sensing*, **28**, 517-535. http://dx.doi.org/10.1080/01431160600821127

[10] Doerffer, R. and Fischer, J. (1994) Concentrations of Chlorophyll, Suspended Matter and Gelbstoff in Case II Waters Derived from Satellite CZCS Data with Inverse Modeling Methods. *Journal of Geophysical Research-Oceans*, **99**, 7457-7466. http://dx.doi.org/10.1029/93JC02523

[11] Lindell, T., Pierson, D., Premazzi, G. and Zilioli, E. (1999) Manual for Monitoring European Lakes Using Remote Sensing Techniques EUR Report, Vol. 18665, Office for Official Publications of the European Communities (EN), Luxembourg.

[12] Härmä, P., Vepsäläinen, J., Hannonen, T., Pyhälahti, T., Kämäri, J., Kallio, K., *et al.* (2001) Detection of Water Quality Using Simulated Satellite Data and Semiempirical Algorithms in Finland. *Science of the Total Environment*, **268**, 107-121. http://dx.doi.org/10.1016/S0048-9697(00)00688-4

[13] Brando, V.E. and Dekker, A.G. (2003) Satellite Hyperspectral Remote Sensing for Estimating Estuarine and Coastal Water Quality. *IEEE Transactions on Geoscience and Remote Sensing*, **41**, 1378-1387. http://dx.doi.org/10.1109/TGRS.2003.812907

[14] Schroeder, Th. and Schaale, M. (2005) Brief Documentation of the FUB/WeW WATER Processor—A Plug-In for MERIS/(A)ATSR Toolbox (BEAM). http://www.brockmann-consult.de/beam/plugins.html

[15] IOCCG (2000) Remote Sensing of Ocean Colour in Coastal, and Other Optically-Complex, Waters. In: Sathyendranath, S., Ed., *Reports of the International Ocean-Colour Coordinating Group*, No. 3, IOCCG, Dartmouth.

[16] Sathyendranath, S. and Platt, T. (1997) Analytic Model of Ocean Color. *Applied Optics*, **36**, 2620-2629. http://dx.doi.org/10.1364/AO.36.002620

[17] Gordon, H.R., Brown, O.B., Evans, R.H., Brown, J.W., Smith, R.C., Baker, K.S., *et al.* (1988) A Semianalytic Radiance Model of Ocean Color. *Journal of Geophysical Research*, **93**, 10909-10924. http://dx.doi.org/10.1029/JD093iD09p10909

[18] Walker, R.E. (1994) Marine Light Field Statistics. Wiley, New York.

[19] Arfi, R., Ba, N., Bouvy, M., Corbin, C., Diop, Y., Ka, S., Lebihan, F., Mboup, M., Ndour, E.H., Pagano, M. and Sané, S. (2003) Lac de Guiers (Sénégal). Conditions environnementales et communautés planctoniques. Document Centre IRD Dakar, 77 p.

[20] Traoré, O. (1995) Etude des échanges hydrogéologiques entre les eaux du lac de Guiers et la nappe alluviale superficiel sous-jascente (Sénégal) Mem. DEA Institut des sciences de l'environnement, 107 p.

[21] Thiam, A. and Ouattara, M. (1997) Un macrophyte en voie d'envahissement du lac de Guiers (Sénégal): *Potamogeton schweinfurthii* A. Bennett (Potamogetonaceae). *Journal de Botanique, SociétéBotanique de France*, **4**, 71-78.

[22] Cogels, F.X., Coly, A. and Niang, A. (1997) Impact of Dam Construction on the Hydrological Regime and Quality of a Sahelian Lake in the River Senegal Basin. *Regulated Rivers: Research & Management*, **13**, 27-41.

[23] Bamba, S.B. (1985) Première approche de la physico-chimie des eaux intertitielles des sédiments du Lac Guiers (Sénégal). Mémoire de DEA, UCAD, Faculté des Sciences, 60 p.

[24] Sane, S. (2006) Contrôle environnemental de la production primaire du lac de Guiers au Nord du Sénégal. Thèse doct. 3e cycle UCAD, Dakar, 187 p.

[25] Ba, N. (2006) La communauté phytoplanctonique du lac de Guiers (senegal): Types d'associations fonctionnelles et approches expérimentales des facteurs de régulation. Thèse doct. 3e cycle UCAD, Dakar, 144 p.

[26] Carl Bro International (1999) Etude bathymétrique et limnologique du lac de Guiers. Rapport de synthèse Hydroconsult international, SGPRE, 119 p.

[27] Rast, M. (1999) The ESA Medium Resolution Imaging Spectrometer MERIS—A Review of the Instrument and Its Mission. *International Journal of Remote Sensing*, **20**, 1679-1680. http://dx.doi.org/10.1080/014311699212416

[28] Diop, S., Wade, S. and Tijani, M.N. (2008) Analysis of MERIS Data for Lake Guiers' (SENEGAL) Water Quality Assessment-Preliminary Results. *Proceedings of the 2nd MERIS/(A)ATSR User Workshop*, Frascati, 22-26 September 2008, 8 p.
https://earth.esa.int/web/guest/missions/esa-operational-eo-missions/envisat/content/-/asset_publisher/V1xF/content/2nd-meris-a-atsr-user-workshop-6094

[29] Fomferra, N. and Brockmann, C. (2005) Beam—The ENVISAT MERIS and AATSR Toolbox. *Proceedings of the MERIS/(A)ATSR User Workshop*, Frascati, 26-30 September 2005.

[30] Binding, C.E., Jerome, J.H., Bukata, R.P. and Booty, W.G. (2007) Spectral Absorption Properties of Dissolved and Particulate Matter in Lake Erie. *Remote Sensing of Environment*, **112**, 1702-1711. http://dx.doi.org/10.1016/j.rse.2007.08.017

[31] Brock, T.C.M., Bos, A.R., Crum, S.J.H. and Gylstra, R. (1995) The Model Ecosystem Approach in Ecotoxicology as Illustrated with a Study on the Fate and Effects of an Insecticide in Stagnant Freshwater Microcosms. In: Hock, B. and Niessner, R., Eds., *Immunochemical Detection of Pesticides and Their Metabolites in the Water Cycle* (*DFG Research Report*), VCH, Weinheim, Basel, Cambridge, New York and Tokyo, 167-185.

[32] Thomann, R.V. and Mueller, J.A. (1987) Principles of Surface Water Quality Modeling and Control. Harper-Collins, New York, 644 p.

[33] Lung, W.S. (1996) Fate and Transport Modeling Using a Numerical Tracer. *Water Resources Research*, **32**, 171-178. http://dx.doi.org/10.1029/95WR02918

14

The Challenges of Water Pollution, Threat to Public Health, Flaws of Water Laws and Policies in Pakistan

Azra Jabeen[1]*, Xisheng Huang[1], Muhammad Aamir[2]

[1]School of Law, Chongqing University, Chongqing, China
[2]College of Materials Science and Engineering, Chongqing, China
Email: *azra.badar@gmail.com

Abstract

In an era of unprecedented urbanization, population and industrial growth pressure is serious threat for the water management in Pakistan in present days. Water pollution from raw sewage, industrial wastes, and agricultural runoff limited natural fresh water resources in the country. Human health is facing serious problems due to deteriorating drinking water quality. Current review paper provides an insight to the water quality problems in Pakistan with an attempt to emphasize the challenges of water laws enforcement. Although Pakistan has developed many water laws the state of implementation is dominant, intermediate pollution crises are still remaining. We could come to the conclusion that strictly enforcement is compulsory for water environment regulations in Pakistan. Moreover, it is necessary to establish a reliable risk assessment system for water quality, human health and ecological safety.

Keywords

Water Pollution, Population, Urbanization, Public Health, Contamination, Industrial and Agricultural Pollution, Water Laws and Policies

1. Introduction

Relationships between water and other development-related sectors such as population, energy, food, and environment, and the interactions among them require analysis, as they together will determine future food security and pollution reduction [1].

---

*Corresponding author.

Everyday two million tons of sewage industrial and agricultural waste is discharged into the world's water. FAO [2] showed that pressure on water resources due to irrigation would be increased by 2050. The UN estimates that the amount of waste water produced annually is about 1500 km^3, six times more water than exists in all the rivers of the world [3].

The world's 1.1 billion people lack access to basic water supplies and half of the developing world's population suffers from diseases due to water contamination [4]. Approximately 2.5 million people of the world live without improved sanitation which is most significant form of water pollution [5]. Unsafe or inadequate water cause approximately 3.1 percent of all deaths worldwide, and 3.7 percent of DALYS worldwide. Infectious diseases such as waterborne diseases are the number one killer of children under five years old [6].

Unsafe water causes 4 million diarrheal diseases and 2.2 million deaths in each year [7] 1.3 million People die of *malaria* [8], 160 million people are infected with *schistosomiasis* (A disease in sub-Saharan Africa) each year. 500 million people are at risk from *trachoma* and 1.5 million cases of clinical *hepatitis A* every year [9]. More than 5 million people die each year from water related diseases [10]. In Pakistan, water remains a critical resource for sustained well being of its citizens. The water availability in Pakistan is continuously declining, both in total amount of water and the per capita water availability. **Table 1** show that in 1951, when population was 34 million, per capita availability of water was 5000 cubic meters, which has now decreased to 1109 cubic meters, reached water scarcity level of 1000 cubic meters [11]. And alarming prediction is less than five hundred cubic meters per capita per year by 2020.

The water shortages and increasing competition for multiple uses of water has adversely affected the quality of water. Consequently, most of the reported health problems are directly or indirectly related to polluted water in the country [12]. A vast majority of Pakistan's population does not have to access to clean, portable and safe drinking water [13]. Water sources like surface and ground water are contaminated with Bacteriological, (Arsenic Nitrates and Fluoride) Contaminations; municipal, agricultural and industrial waste throughout the country. Despite high population of the country, water sources improved from 85% in 1990 to 92% in 2010, but water from these sources areas are not save to drink. However the water sector still has challenges and quality of services is poor. Drinking water quality is deteriorating continually due to biological contamination from human waste, chemical pollutants from industries and agricultural inputs. Piped water also gets contaminated because pipes are laid very close to sewerage lines or open drains and cause many serious water borne diseases. It was found that 45% of infant deaths have been attributed to diarrhea and about 60% to overall infectious water borne diseases in Pakistan. According to the World Health Organization (WHO) 25% - 30% of the diseases are gastrointestinal in nature [13].

In the provinces of Pakistan, Sindh 24%, NWFP 46% and Blochishtan 72% of population rely on unsafe water sources [14]. Pakistan is ranking 17 among countries facing watershortage [15] and 79 percent water is unsafe for drinking [16]. According PSLM Survey of 2010-11, [17] the main source of drinking water was: 32% tap water, 28% hand pump, 27% motor pump, 4% dug well and 9% others. Pakistan's 96% of urban and 89% of rural population have access in water pollution in broad definition and 57% of urban and 15% of rural population access houses connection water pollution.

Table 1. Per capita water availability.

Years	Population (million)	Per capita availability (m^3)
1951	34	5000
1961	46	3950
1971	65	2700
1981	84	2100
1991	115	1600
2000	148	1200
2013	178	1105
2015	190	1000
2025	267	659

Sources: Hassan Ahmed Khan ppt on water pollution in Pakistan. 30 June 2010 that was consists on prediction but after five years now situation is different. So here are some changes according to current situation.

The reasons why the Pakistan continuously fails in protecting water sources from scarcity and pollution have many aspects. The main reasons including unclear strategies of laws and polices enforcement and no coordination between relevant departments, lack of accountability and transparency of water based regulatory authorities, corrupt and incompetent governmental functionaries, lack of public participation. 1) There should be a strict legislation system that would serve as a base for monitoring the implementation of water pollution laws and polices; 2) For environmental legislation to "work" it must not only be well designed but also efficiently and effectively enforced. Strategies must be developed; 3) Pakistan's decision makers and those in power are not directly affected by water insecurities and have little incentive to change the system or make water management reform a priority

1.1. Population and Environmental Issues

There are alternative views on population-environment linkages [18]. Population and environment are closely intertwined in acomplex and dynamic relationship. The relationship between population and environmentis mediated by a number of socioeconomic, cultural, political, and developmental variables whose relative significance varies considerably from one context to another. Over the past three to four decades, some economists, biologists, and environmentalists have been debating the role of population in environmental degradation.

At the time of independence in 1947, 32.5 million people lived in Pakistan. By 2006-2007, the population is estimated to have reached 156.77 million. Thus, in roughlythree generations, Pakistan's population has increased by 124.27 million or has grown at an average rate of 2.6 percent per annum. The present population of the country is 188 million will increase to nearly 190 million by 2016, to over 220 million by 2020 and to almost 275 million by 2050, as Pakistan retains its position as the sixth most populous country in the world [19]. **Table 2** shows the population growth rate, urbanization, and industrialization in Pakistan from 1980 to 2025.

Figure 1 shows that from the year 2001 to 2011 if we compare Pakistan to 10 most populous countries, it ranks first in increasing growth rate annually.

Table 2. Population, urbanization, and industrialization in Pakistan, 1980-2025.

Year	Population (million)	Population growth rate (%)	Urbanization (% of population)	Industrialization (share of manufacturing in real GDP in %)	Water demand for domestic use (MAF)
1980	84.9	3.0	24.1	13.8	
1990	110.8	2.6	34.7	17.4	4.1
2000	140.5	2.1	47.5	23.0	5.2
2010	177.1	2.1	36	25.4	
2025	228.8	2.4	49.3		9.7

Sources: Population Reference Bureau (2004); Ministry of Finance & Economic Affairs (2004, 2001); Federal Bureau of Statistics (1991); Kahlown and Majeed (2002).

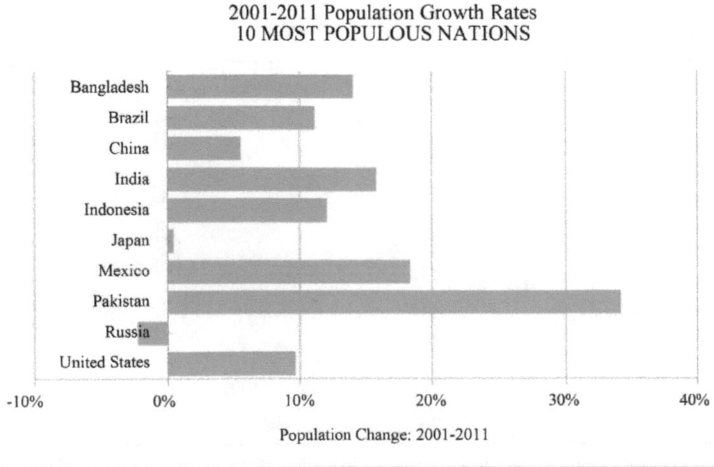

Figure 1. Population growth rates.

1.2. Municipal, Industrial and Agriculture Water Pollution

The pressure on water resources caused by industrial growth also merits discussion due to their significant contribution to water pollution problems. It has been estimated that around 2000 million gallon of sewage is being discharge to surface water bodies everyday in Pakistan [20]. According to Sial [21] in Pakistan out of 6634 registered industries 1228 are considered to be highly polluting. Industrial units including textile, chemical, food processing, pulp and paper, poultry, dairy, plastic, paint, pesticides, leather, tanneries and pharmaceuticals directly discharged their waste into the canal system contaminating ground water level as well [22]. In Pakistan, only 1% of wastewater is treated by industries before being discharged directly into rivers and drains. In NWFP, 80,000 m^3 of industrial effluents containing a very high level of pollutants are discharged every day into the river Kabul. In Karachi, Sindh Industrial Trading Estate (SITE) and Korangi Industrial and Trading Estate (KITE), two of the biggest industrial estates in Pakistan, there is no effluent treatment plant and the waste containing hazardous materials, heavy metals, oil etc is discharged into rivers. In Multan, a fertilizer factory discharges its waste untreated to cultivated land causing death of livestock and increasing health risk to humans [23]. In Lahore, only 3 out of some 100 industries using hazardous chemicals treat their wastewater. All the big cities like Karachi dumped its 600 million-tones sewage daily into the sea. Lahore dumped about 200 million tones liquid and 100 million tones solid wastes into the river Ravi. Due to open dumping of industrial/municipal wastes, the underground quality of water is deteriorating. The discharge of wastewater from domestic, municipal and industrial sectors directly into water bodies without proper treatment is major cause of surface and groundwater pollution in Pakistan Water pollution also influences the agriculture of Pakistan with excessive use of fertilizers and pesticides which are key pollutants as they dissolve in water and seep to the underground water bodies becoming the source of salinity and water logging at irrigated land decreases the fertility of soil, costing Pakistan 0.9 per cent of the GDP. The contribution of agricultural drainage to the overall contamination of the water resources exists but is marginal compared to the industrial and domestic pollution. For example, in Sindh, the pollution of water due to irrigation is only 3.21% of the total Pollution [24].

1.3. Bacteriological Contaminations and Drinking Water Quality

Arsenic Toxicity Investigations revealed the presence of excessive arsenic in many cities of Punjab and Sindh provinces was found to be 50 ppb five times higher than the prescribed limit of 10 ppb by WHO [25].

Detailed data analysis has identified 4 major water quality tribulations in drinking water sources of Pakistan i.e. bacteriological (68%), arsenic (24%), nitrate (13%) and fluoride (5%). The five years trend analysis has revealed that out of a total 357, only 45 water sources (13%) were found "Safe" and the remaining 312 (87%) were "Unsafe" for drinking purpose. The water quality monitoring (2001-2010) conducted in rural and urban areas of the country revealed that access to save drinking water is only 15 percent in urban and 18 percent in rural areas.

Table 3 shows bacterial contamination level in 23 cities of Pakistan from 2002 to 2006 which increased every year.

On the behalf of PCRWR survey published in 2012, 88% of the functional water supply schemes in Pakistan provide water that is unsafe for drinking because of microbiological contamination [26]. According to an official government document [27] increased arsenic, nitrate and fluoride contamination was detected in drinking water in various localities in Pakistan. A survey of drinking water samples in Karachi in 2007/08 found that, of 216 ground and surface water samples collected, 86% had lead levels higher than the WHO maximum acceptable concentration of 10 parts per billion (ppb). This mean lead concentration was 146 ppb in untreated ground water and 77 ppb in treated tap water [28]. Newspapers reported that a citizen of Karachi submitted a court petition asking the Karachi Water and Sewerage Board (KWSB) to fulfill its duty to provide clean water. In October 2012, the Sindh High Court issued a notice to the Board asking it to comment on the petition [29].

The Nation-wide Assessment Survey of more than 10,000 water supply schemes (1808 urban and 8320 rural water supply schemes) carried out by the PCRWR revealed that 72 percent schemes are operational and only 23 percent in urban and 14 percent in rural areas water supply schemes are supplying safe drinking water [30].

1.4. Water Pollution and Health Problems

In Pakistan contamination of drinking water with industrial wastes and municipal sewage coupled with lack of

Table 3. The bacterial contamination level in major cities of pakistan.

Name of city	Contamination level in 2002 %	Contamination level in 2006 %
Islamabad	40	74
Faisalabad	38	79
Bahawalpur	52	75
Gujranwala	29	71
Gujrat	56	100
Kasur	40	50
Lahore	37	63
Multan	31	87
Rawalpindi	53	87
Sheikhupura	27	55
Sialkot	40	70
Sargodha	75	92
Khuzdar	62	100
Loralai	73	100
Quetta	48	68
Ziarat		100
Mangora	40	70
Mardan	75	83
Peshawar	31	77
Abbottabad	55	73
Hyderabad	73	100
Karachi	61	100
Sukkur	67	83

Sources: PCRWR report 2002-2006.

water disinfection practices and quality monitoring at treatment plants is the main cause of the prevalence of waterborne diseases [31].

Around 62 percent of Pakistan's urban and 84 percent of its rural population does not treat their water, resulting in 100 million cases of diarrheal diseases registered in hospitals, with 40 percent of deaths attributed to drinking polluted water. The diarrhea which is a water-linked disease, accounts for 14% of illnesses, One third of under-five deaths in children [32] and for 7% of all diseases in people of all ages in Pakistan [33]. Diarrheal rate in Pakistan is the second highest amongst 31 Asian countries. Although in the developed countries typhoid fever has been almost eliminated, in developing countries like Pakistan it is still a common disease and a major cause of morbidity and mortality due to lack of sewage and water treatment facilities [34]. Lack of effective prevention and control measures contribute in worsening the situation [35].

According to estimation 250,000 child deaths occur each year in Pakistan due to water-borne diseases [30] and more than 1.6 million DALYs are lost annually as a result of death and ailment due to diarrhea and almost 90,000 as a result of typhoid. Inadequate quantity and quality of potable water is associated with a host of illnesses. 20 - 40 percent of the hospital beds in Pakistan are occupied by patients suffering from water-related diseases [36].

2. Legal Framework and Its Enforcement

2.1. Legal Protection of Water from Pollution

In 1997 a regulatory framework known as PEPA Act 1997 was approved to regulate and monitor issues regarding environmental protection in the country [37].

Table 4 makes it clear that environmental legislation to sanction water pollution does exist in Pakistan. Under federal legislation, the relevant provisions relating to the human right to water, including the prevention of water pollution, provisions of the PEPA Act, 1997 relating to the disposal of wastes and effluents and Art. 20 relating to drinking water, of the amended Factories Act of 1934 [38]. The Pakistan Penal Code1860 contains a criminal penalty for polluting the water of any public spring or reservoir [39]. Another relevant piece of legislation is PCRWR Act 2007, which set up the Pakistan Council of Research in Water Resources, which is primarily entrusted with improving the technology needed to advance, as well as to conserve existing water resources. This Body is also required to provide recommendations to the government, regarding the quality of water that needs to be maintained and how existing water sources may be utilized and conserved [40] furthermore, various water and sanitation based policies and guidelines have been approved by the national government. The IRSA Act 1992 implements the Water Accord which apportions the balance of river supplies, including flood surpluses and future storages among the provinces. The WAPDA Act 1958, Water User Ordinances 1982. The PIDA Acts

Table 4. Key water-related legislation in pakistan.

Date	Legislation	Implementation	Key feature
1860	Pakistan Penal Code	Federal Penalizes	Water pollution as a public health issue
1873	Canal and Drainage Act	Federal and provincial	Governs irrigation water use
1882	Easement Act	Federal	Grants and limits rights for water pollution
1883	Land Improvement Loans Act	Provincial	Provides loans for water distribu-tion, drainage, and reclamation
1905	Punjab Minor Canals Act	Provincial	Governs irrigation water use
1927	Forests Act	Federal and Provincial	Governs disposal of waste and effluent
1934	Factories Act	Federal	Penalizes pollution of water in forests
1949	Karachi Joint Water Board Ordinance	Municipal	Prohibits pollution of water supply; first water law at municipal level
1952	Punjab Development of Damaged Areas Act	Provincial	Allows government to construct sewage and drainage in "damaged areas"
1958	West Pakistan Water and Power Development Authority Act	Provincial and Federal	Establishes what is today the Water and Power Development Authority
1960	Indus Waters Treaty	International	Governs sharing of Indus River waters between Pakistan and India
1976	Territorial Waters and Maritime Zones Act	Federal, International	Declares maritime territory and boundaries
1980	Sindh Fisheries Ordinance	Provincial	Prohibits dumping of pollutants in water
1981	On-Farm Water Manage-ment and Water Users' Association Ordinance	Federal	Provides resources for improved irrigation water managemen
1991	Indus River System Accord	Provincial	Governs water sharing between provinces
1997	Provincial Irrigation and Drainage Authority Acts	Provincial	Implements irrigation reforms
1997	Environmental Protection Act	Federal	Governs protection, conservation, rehabilitation, pollution, and improvement of environment
2009	National Drinking Water	Federal	Provides institutional framework and guidelines for provinces to assure quality and supply of drinking water
2010	18th Amendment to the Constitution of Pakistan	Federal	Devolves Ministry of Environment to provinces, establishes forums for interprovincial dialogue
2011	Punjab Environmental Protection (Amendment) Bill	Provincial	Establishes Punjab Ministry of Environment

Source: Jawad Hassan, Environmental Laws of Pakistan (Lahore: Book Biz, 2006).

1997, The Sindh Irrigation Act 1879, Provincial legislation such as the BGWRA Ordinance, IX Of 1978, established regulatory and supervisory functions for the Provincial Water Board and a Water Committee to overlook the implementation of the policies of the Water Board [41]. Other laws related to pollution prevention of water bodies include the Canal and Drainage Act (1873) and the Punjab Minor Canals Act (1905), which prohibits the corrupting or fouling of canal water; Sindh Fisheries Ordinance (1980), which prohibits the discharge of untreated sewage and industrial waste into water, and The Greater Lahore Water Supply Sewerage and Drainage Ordinance (1967) all these legislation related to water rights. However, under the PLGO (2001), [42] a number of provincial functions including water management and sanitation have been entrusted to the TMAs. The CDG and the TMAs [43] are also responsible for the enforcement of punishment for offences, relating to the contamination or pollution of water, failure on the part of industries to dispose of hazardous waste, or offences relating to the provision of contaminated water for human consumption [43] Other forms of offences such as failure to stop leakage of drain pipes, the obstruction of water pipes etc. have been made punishable by the issuance of tickets rather than through court and are the responsibility of the Tehsil/Town Officer.

Furthermore, the National Government has approved various water policies and guidelines. In November 2002 national standards for drinking water quality were introduced. Similarly other policies including National Environment Policy 2005, [44] National Sanitation Policy 2006 [45] and National Drinking Water Policy 2009 [46] have been approved. The National Environment Policy, 2005, provides a framework for various environmental issues, particularly the pollution of fresh-water bodies. It recognizes the need to meet international obligations effectively and in line with national objectives [47] and concerns regarding public health and environment. In addressing water supply and management, it lists a number of guidelines by which the government can ensure sustainable access to safe water resources [48].

Under the National Drinking Water Policy 2009, water was the basic human right of every citizen [49]. The policy aims at providing safe drinking water to the entire Pakistani population by 2025, including the poor and vulnerable, at an affordable cost [50].

As per policy mandates that the safe drinking water be accessible to both urban and rural areas. The policy declares that various forms of legislation are to be enacted to ensure the implementation of these measures, including the Pakistan Safe Drinking Water Act [51]. Policy's objectives briefly described in **Table 5**.

2.2. Weaknesses in the Enforcement Status

Although Pakistan has comprehensive national laws and policies related to water pollution control and institutional framework for environmental management, yet there are significant weaknesses in the current administrative and implementation capacity. Water disputes in Pakistan are chronic. The perennial irrigation water shortages create conflict between provinces. The lack of water laws that define water rights often pit users against each other. Most of the water-related legislation and regulations, dating back to colonial rule as evident in **Table 3**. The Pakistan Penal Code 1860 and the Factories Act 1934 are quite old and very weakly enforced, defining penalties which are no longer effective. Although IRSA 1992, relatively effective in distributing water to the provinces in the past, the authority has recently indicated that it expects acute water shortages for irrigation coordination and communication between federal, provincial and local administrative entities is curtailed. Unclear policies definition plans and targets is another Impuissance. Pakistan's policy approach, which is oriented in supply-side interventions, is also lacking. Over the past decade, Islamabad drafted a number of policies and strategies to address the country's various water challenges. These include the Environment Policy (2005), National Sanitation Policy (2006), the Drinking Water and Sanitation Policy (2009), but have yet to be adopted in real sense.

Although ordinances, acts and policies have been approved from time to time, therefore clear strategies are so far from their implementation. As a result, after appropriate and necessary administrative capacity on paper, its effectiveness is seriously curtailed in practice. The industries do not follow the national standards for pollutants in their waste effluents. Industrial effluent is to be regulated by environment protection agencies through self-monitoring and reporting programmes under PEPA Act, but proverbially, enforcement is lax [52]. The municipal wastewater treatment facilities in Karachi and in Islamabad have not been able to achieve the performance standards set under the NEQS. Groundwater or surface water can be acquired at a lower cost in Punjab and NWFP where supply is generally abundant, and the only option available is disposal of the treated effluent into the open drains. This option can be exercised, but will require stricter monitoring particularly when the canals are closed or in low flow conditions.

Table 5. Objectives of national drinking water policy 2009.

Targeting strategy	Provide water to un-served or under-served areas where the access to clean drinking water have walk of 0.5 KM
Legislative strategy	Encourage the participation of private sector. community participation, public-private partnership and role of NGOs
Protection of water resources	Protection of surface and ground water sources in urban and rural areas
Institutional strategy	Strengthen the institutions and their services of providing drinking water at federal, provincial and local government level
Technical strategy	Technical assistance to the provincial and local agencies, support clean drinking water initiative project by the federal government and provincial water filtration plants at tehsil and union council level
Operation and maintenance (O&M) strategy	WASA and TMA are responsible for the O&M in urban areas. TMA is only responsible for O&M in rural areas
Drinking water quality standards	Using the WHOs drinking water quality guidelines both in urban and rural areas
Water quality monitoring and surveillance	PQCA and PCRWR ensure the water quality standard through surveillance of water quality from different sources
Gender strategy	Female participation in decision making at the district, tehsil and union council tiers
Communication and dissemination strategy	*Disseminating information* on drinking water quality standards through articles in the national press, leaflets, newsletters and spreading information to schools, through NGOs, civil societies, and citizen community boards (CCBs).
Financial strategy	Provision of water supply and sanitation services at affordable rates is Promised in the *financial strategy*. funds are given to CCBs at local level for drinking water schemes
Monitoring and evaluation strategy (M&E),	Ministry of environment will be responsible for reporting of *State of Drinking Water in Pakistan*, whereas, the local government department and WASAs will be responsible for monitoring the coverage of drinking water supply in rural and urban areas respectively.
Research strategy	Pilot tests for new approaches and innovative ideas in the drinking water sector, especially those which help to improve access, efficiency, effectiveness and sustainability.

Government has introduced different programs to control water pollution, but unfortunately no one is implemented appropriately due to weak law enforcement and the problems remained the same. Policy is not the only challenge, Citizen Education is also necessary to overcome the public's apathy toward water pollution. But citizen education requires support from the government in terms of legislation, conservation strategies, and law enforcement, which is currently nonexistent. There is a deep mistrust among stakeholders and decision makers, partly explaining the stalemate in passing meaningful water policies and strategies. The lack of consensus on water sector priorities not only creates a vacuum for improved resource management but also leaves the security around water that much more volatile. Thus it is crucial to correct these misperceptions if there is to be any significant shift in water policy and management. Water stakeholders and policymakers would be better served using their collective skills, knowledge, and expertise to develop an effective strategy that tackles Pakistan's precarious water environment and off sets any violence over water resources.

3. Conclusions

Growing population, poor management services, water pollution, lack of public awareness and weak enforcement of environmental laws and policies causes great environmental degradation and health problems.

The issues of water quality and quantity are still major problems in Pakistan. Water pollution and its effects over environmental sustainability in Pakistan provoke political instability when other problems and grievances already exist. As shortages become more widespread, it is crucial that the government invest greater political capital to regulate water pollution and provide quality water services to all communities. A number of factors need to be highlighted and addressed in order to improve, protect and maintain the quality of freshwater resources of the country. The Government does not give the priority to the treatment of sewage and industrial effluents. The level of commitment from government authorities to treat waste water and to improve the quality of freshwater is very low. Therefore, there is a need to bring provision of clean water back as a top priority.

There is need to establish legal rules and regulations at the earliest to cover the risks of ground water extraction. Even when there are relevant laws in the country like PEPA 1997, their enforcement is extremely weak and therefore the level of compliance is low particularly in the industrial and housing sector. And policies like NEP 2005, NWP 2006 and NDWP 2009, etc. are in place, there is no clear strategy devised so far to implement them. A clear and practical strategy needs to be defined to implement these policies.

Effective management can only come from domestic reform, and dependence on foreign aid will not render lasting solutions. This study is by no means a complete analysis of the pollution challenges facing Pakistan's water sector. Its goal is to demonstrate how policymakers and water lawyers seem to underestimate the extent of the potential threats to water and its economic future. More data and analysis are needed to understand the extent of each of these challenges and subsequent security threats to pinpoint potential hot spots. In particular, effective laws and policies require greater supply and demand linkages, as opposed to the field's supply-oriented literature. The future of Pakistan's water sector does not have to be ominous. There are great opportunities to address these challenges and avert violent conflict. Ultimately, change starts with political will.

Acknowledgements

Authors are grateful for the support by China Scholarship Council (CSC) and School of Law, Chongqing University for carry of present research study.

References

[1]   Valipour, M. (2015) Assessment of Important Factors for Water Resources Management in European Agriculture. *Journal of Water Resource and Hydraulic Engineering*, **4**, 171-180.

[2]   FAO (2011) Climate Change, Water and Food Security (FAO Water Reports Paper No. 36) [Internet]. FAO, Rome. http://www.fao.org/docrep/014/i2096e/i2096e.pdf

[3]   UNWWAP (United Nations World Water Assessment Programme) (2003) The World Water Development Report 1: Water for People, Water for Life. UNESCO, Paris.

[4]   Erik, B.B. (2004) The Implications of Formulating a Human Right to Water. *Ecology Law Quarterly*, **31**, 957-959.

[5]   UNICEF/WHO (2008) UNICEF and WHO Joint Monitoring Programme for Water Supply and Sanitation. Progress on Drinking Water and Sanitation: Special Focus on Sanitation. UNICEF, New York and WHO, Geneva.

[6]   WHO (2002) Reducing Risks, Promoting Healthy Life. France. http://www.who.int/whr/2002/en/whr02_en.pdf

[7]   WHO, UNICEF (2000) Global Water Supply and Sanitation Assessment 2000 Report.

[8]   WHO (World Health Organization) (2005) Water, Sanitation and Hygiene Links to Health.

[9]   WHO (World Health Organization) (2004) World Health Report: Changing History. Geneva.

[10]   World Water Day (2002) Water for Development. Friday, 22 March 2002 Is World Water Day with the Theme Water for Development.

[11]   The Economist (2012) Going with the Flow. Accessed 7 June 2012. http://www.economist.com/node/21546883

[12]   Ismat Sabir (2012) Water Is Becoming Scarce. Pakistan Observer, 28 November 2012.

[13]   PCRWR (2015) Ministry of Science & Technology. Khyban-E-Johar, H-8/1, Islamabad. www.pcrwr.gov.pk

[14]   Dawn News (2011) No Access to Safe Drinking Water in Pakistan. December 2011, Accessed 08 June 2012. http://dawn.Com/2011/12/28/no-access-to-safe-drinking-water-in-Pakistan/

[15]   The Nation (2009) Pakistan among 17 Countries Facing Water Shortage. September 2009, Accessed 1 March 2012. http://www.nation.com.pk/pakistannews-newspaper-daily-english-online/karachi/25-Sep-2009/Pak-among-17-countries-facing-water-shortage

[16]   PCRWR (2005) Report on Technical Assessment of Water Supply Schemes. 41-42. 23 November 2011 (Accessed 3 January 2012). http://www.pcrwr.gov.pk/Publications/Final/Punjab/Report/Volume http://www.pakistantoday.com.pk/2012/12/24/city/islamabad/pakistans-population-to-increase-to-190-million-by-2015-report/#sthash.Dypb0FJr.dpuf

[17]   PSLM (2010) Percentage Distribution of Households by Source of Drinking Water. TABLE: 4.7, November 2010, Retrieved on 6 November 2012.

[18] Marcoux, A. (1999) Population and Environmental Change: From Linkages to Policy Issues. Sustainable Development Department, Food and Agricultural Organization (FAO) of the United Nations.

[19] Pakistan Day (2012) December 2012.

[20] Pak-SECA (2006) Pakistan Strategic Country Environmental Assessment Report: Rising to the Challenges. Islamabad, Pakistan: South Asia Environment and Social Development Unit.

[21] Sial, R.A., Chaudhary, M.F., Abbas, S.T., Latif, M.I. and Khan, A.G. (2006) Quality of Effluents from Hattar Industrial Estate. *Journal of Zhejiang University SCIENCE B*, **7**, 974-980. http://dx.doi.org/10.1631/jzus.2006.B0974

[22] Raza, A. (2014) Untreated Waste Polluting Underground Water: Report. *The News International*, 26 November 2014.

[23] Shahid, K. (2005) Country Water Resources Assistance Strategy "Drinking Water and Sanitation Sector". Background Paper≠8, Review of Policies and Performance and Future Options for Improving Service Delivery, March 2005.

[24] SOE (2005) Ministry of Environment. Government of Pakistan, draft.

[25] PCRWR (2007) Report.

[26] PCRWR (2012) 79% Water Sources in Punjab Supplying Unsafe Drinking Water. The Tribune, 15 May 2012, Retrieved 7 November 2012.

[27] MPD (2004) Medium Term Development Framework 2005-2010. Section 10: Water and Sanitation. Government of Pakistan, Islamabad, Retrieved 29 May 2008, Section 10.3.

[28] Ul-Haq, N., Arain, M.A., Badar, N., Rasheed, M. and Haque, Z. (2011) Drinking Water: A Major Source of Lead Exposure in Karachi, Pakistan. *Eastern Mediterranean Health Journal*, **17**, 882-886.

[29] The News (2012) KWSB Told to Supply Chlorinated Water. Karachi, November 2012.

[30] Zuhaeb, N. (2012) Contaminated Water Contributes 40% Deaths in Pakistan, UN.

[31] Hashmi, I., Farooq, S. and Qaiser, S. (2009) Chlorination and Water Quality Monitoring within a Public Drinking Water Supply in Rawalpindi Cantt (Westridge and Tench) Area, Pakistan. *Environmental Monitoring and Assessment*, **158**, 339-403. http://dx.doi.org/10.1007/s10661-008-0592-z

[32] Qutub, S.A. (2004) Sanitation and Hygiene in Pakistan. *Proceedings of the National Workshop on Water and Sanitation and Exposition* 2004, National Environmental Consulting (Pvt.) Ltd./Pakistan Institute for Environment Development Action Research (PIEDAR), Islamabad, 10-12 June 2004, 38-45.

[33] Rosemann, N. (2005) Drinking Water Crisis in Pakistan and the Issue of Bottled Water: The Case of Nestlé's "Pure Life". Swiss Coalition of Development Organizations and Actionaid, Pakistan. http://www.alliancesud.ch/en/policy/water/downloads/nestle-pakistan.pdf

[34] Ahmed, H.N., Niaz, M., Amin, M.A., Khan, M.H. and Parhar, A.B. (2006) Typhoid Perforation Stills a Common Problem: Situation in Pakistan in Comparison to Other Countries of Low Human Development. *Journal of Pakistan Medical Association*, **56**, 230-232.

[35] Qasim, M. (2008) Twin Cities: Water-Borne Diseases on the Rise. *The News*, Pakistan, Sunday 20 July 2008.

[36] Khan and Shaheen, R. (2002) State of the Environment Report for Pakistan. Islamabad: SDPI.

[37] Govt of Pakistan (1997) Pakistan Environmental Protection Act, 1997.

[38] Art 14, 20 (1997) Disposal of Wastes and Effluents and Quality Drinking Water. PEPA Act 1997.

[39] PPC (1860) Pakistan Penal Code Chapter XIV: Offences Affecting the Public Health, Safety, Convenience, Decency and Morals 277.

[40] PCRWR (2007) See Section 4—Functions of the Pakistan Council of Research in Water Resources-Act 1 of 2007—Pakistan Council of Research in Water Resources Act, 2007—"...b) Design, develop and evaluate water conservation technologies for irrigation, drinking and industrial water... [a]dvise the government and submit the policies recommendations regarding water quality, development, management, conservation and utilization of water resources...publish scientific papers, reports and periodicals as well as to arrange seminars, workshops and conferences on water related issues; ..."

[41] BGWRA (1978) Ordinance IX (Baluchistan Ground Water Right Administration Ordinance). Functions of WASA in Baluchistan (Section 3), Establishment and Functions of Provincial Water Board.

[42] Punjab Local Government Ordinance (2001).

[43] Punjab Local Government Ordinance (2001). Fourth Schedule—Part D.

[44] MOE-PAK (2005) National Environmental Policy. Govt. of Pakistan: Ministry of Environment.

[45] MOE-PAK (2006) National Sanitation Policy. Govt. of Pakistan: Ministry of Environment.

[46] MOE-PAK (2009) National Drinking Water Policy. Govt. of Pakistan: Ministry of Environment.

[47] NEP: National Environment Policy (2005) Section2.2 (d).

[48] NEP: National Environment Policy (2005) Water Supply and Management. Section 3.1.

[49] NDWP: National Drinking Water Policy (2009) Ministry of Environment, Government of Pakistan. September 2009, Retrieved 7 March 2010, draft.

[50] The Nation (2009) Cabinet Okays National Drinking Water Policy. 29 September 2009, Accessed on 7 March 2010.

[51] NDWP (2009) Pakistan Safe Drinking Water Act Will Be Enacted to Ensure Compliance with the National Drinking Water Quality Standards. Section 6.12.

[52] Alam, A.R. (2012) The Pakistan Water Quality Crises. *The Express Tribune*, 15 March 2012.

Permissions

All chapters in this book were first published in JWARP, by Scientific Research Publishing; hereby published with permission under the Creative Commons Attribution License or equivalent. Every chapter published in this book has been scrutinized by our experts. Their significance has been extensively debated. The topics covered herein carry significant findings which will fuel the growth of the discipline. They may even be implemented as practical applications or may be referred to as a beginning point for another development.

The contributors of this book come from diverse backgrounds, making this book a truly international effort. This book will bring forth new frontiers with its revolutionizing research information and detailed analysis of the nascent developments around the world.

We would like to thank all the contributing authors for lending their expertise to make the book truly unique. They have played a crucial role in the development of this book. Without their invaluable contributions this book wouldn't have been possible. They have made vital efforts to compile up to date information on the varied aspects of this subject to make this book a valuable addition to the collection of many professionals and students.

This book was conceptualized with the vision of imparting up-to-date information and advanced data in this field. To ensure the same, a matchless editorial board was set up. Every individual on the board went through rigorous rounds of assessment to prove their worth. After which they invested a large part of their time researching and compiling the most relevant data for our readers.

The editorial board has been involved in producing this book since its inception. They have spent rigorous hours researching and exploring the diverse topics which have resulted in the successful publishing of this book. They have passed on their knowledge of decades through this book. To expedite this challenging task, the publisher supported the team at every step. A small team of assistant editors was also appointed to further simplify the editing procedure and attain best results for the readers.

Apart from the editorial board, the designing team has also invested a significant amount of their time in understanding the subject and creating the most relevant covers. They scrutinized every image to scout for the most suitable representation of the subject and create an appropriate cover for the book.

The publishing team has been an ardent support to the editorial, designing and production team. Their endless efforts to recruit the best for this project, has resulted in the accomplishment of this book. They are a veteran in the field of academics and their pool of knowledge is as vast as their experience in printing. Their expertise and guidance has proved useful at every step. Their uncompromising quality standards have made this book an exceptional effort. Their encouragement from time to time has been an inspiration for everyone.

The publisher and the editorial board hope that this book will prove to be a valuable piece of knowledge for researchers, students, practitioners and scholars across the globe.

List of Contributors

Deborah Leslie, Kathleen Welch and William Berry Lyons
School of Earth Sciences, The Ohio State University, Columbus, USA

Eliene Lopes de Souza, Roseli de Almeida and Marcio Cabral
Institute of Geosciences, Federal University of Pará (UFPA), Belém, Brazil

Paulo Galvão
Institute of Geosciences, University of São Paulo, São Paulo, Brazil

Cleane Pinheiro
Institute of Environment and Spatial Planning of the State of Amapá (IMAP), Macapá, Brazil

Marcus Baessa
Leopoldo Américo Miguez de Mello Center for Research and Development (CENPES), Rio de Janeiro, Brazil

Emma E. Ezenwaji and Joseph E. Ogbuozobe
Department of Geography and Meteorology, Nnamdi Azikiwe University, Awka, Nigeria

Bede M. Eduputa
Department of Environmental Management, Nnamdi Azikiwe University, Awka, Nigeria

Hussein Mahmood Shukri Hussein
Biotechnology Research Center, Al-Nahreen University, Baghdad, Iraq

Adelere E. Adeniran
Civil & Environmental Engineering Department, University of Lagos, Lagos, Nigeria

Olufemi A. Bamiro
Mechanical Engineering Department, University of Ibadan, Ibadan, Nigeria

Nianqing Zhou
Department of Hydraulic Engineering, Tongji University, Shanghai, China

Telesphore Habiyakare
Department of Hydraulic Engineering, Tongji University, Shanghai, China
College of Science and Technology, University of Rwanda, Kigali, Rwanda

Lei Qiu
State Key Laboratory of Hydrology-Water Resources and Hydraulic Engineering, Hohai University, Nanjing, China
Research Institute of Management Science, Business School, Hohai University, Nanjing, China

UNESCO-IHE Institute for Water Education, Delft, The Netherlands

Huimin Wang
State Key Laboratory of Hydrology-Water Resources and Hydraulic Engineering, Hohai University, Nanjing, China
Research Institute of Management Science, Business School, Hohai University, Nanjing, China

Meine Pieter Van Dijk
UNESCO-IHE Institute for Water Education, Delft, The Netherlands

Devendra Amatya and Carl Trettin
Center for Forested Wetlands Research, USDA Forest Service, Charleston, USA

Timothy Callahan
Department of Geology and Environmental Geosciences, College of Charleston, Charleston, USA

William Hansen
Formerly with Francis Marion and Sumter National Forests, USDA Forest Service, Columbia, USA

Artur Radecki-Pawlik
Agricultural University of Krakow, Krakow, Poland

Patrick Meire
Department of Biology, University of Antwerp, Antwerp, Belgium

Yericho Berhanu Meshesha
School of Natural Resources and Environmental Studies, Hawassa University, Hawassa, Ethiopia

Belay Simane Birhanu
College of Development Studies, Addis Ababa University, Addis Ababa, Ethiopia

Yun-Ya Yang and Gurpal S. Toor
Soil and Water Quality Laboratory, Gulf Coast Research and Education Center, Institute of Food and Agricultural Sciences, University of Florida, Wimauma, FL, USA

Akira Hiratsuka
Osaka Sangyo University, Osaka, Japan

Yugo Tomonaga
National Museum of Ethnology, Osaka, Japan

Yoshiro Yasuda
University of Hyogo, Kobe, Japan

Seybatou Diop and Souléye Wade
Institute of Earth Sciences, Faculty of Sciences and Techniques, Cheikh Anta Diop University of Dakar, Dakar, Senegal

Moshood N. Tijani
Department of Geology, University of Ibadan, Ibadan, Nigeria

Azra Jabeen and Xisheng Huang
School of Law, Chongqing University, Chongqing, China

Muhammad Aamir
College of Materials Science and Engineering, Chongqing, China